单片机实验及实训

总主编 胡学钢
主　编 刘乐群　黄大君

图书在版编目(CIP)数据

单片机实验及实训/刘乐群,黄大君主编.合肥:安徽大学出版社,2016.5
工程应用型院校计算机系列教材
ISBN 978-7-5664-1063-4

Ⅰ.①单… Ⅱ.①刘… ②黄… Ⅲ.①单片微型计算机－高等学校－教材 Ⅳ.①TP368.1

中国版本图书馆 CIP 数据核字(2016)第 040665 号

单片机实验及实训

刘乐群　黄大君　主　编

出版发行：	北京师范大学出版集团 安 徽 大 学 出 版 社 (安徽省合肥市肥西路 3 号 邮编 230039) www.bnupg.com.cn www.ahupress.com.cn
印　　刷：	合肥现代印务有限公司
经　　销：	全国新华书店
开　　本：	184mm×260mm
印　　张：	12
字　　数：	292 千字
版　　次：	2016 年 5 月第 1 版
印　　次：	2016 年 5 月第 1 次印刷
定　　价：	26.00 元

ISBN 978-7-5664-1063-4

策划编辑：李　梅　蒋　芳		装帧设计：李　军
责任编辑：蒋　芳		美术编辑：李　军
责任校对：程中业		责任印制：李　军

版权所有　侵权必究
反盗版、侵权举报电话：0551－65106311
外埠邮购电话：0551－65107716
本书如有印装质量问题，请与印制管理部联系调换。
印制管理部电话：0551－65106311

编写说明

计算机科学与技术的迅速发展,促进了许多相关学科领域以及应用分支的发展,同时也带动了各种技术和方法、系统与环境、产品以及思维方式等的发展。由此而进一步激发了对各种不同类型人才的需求。按照教育部计算机科学与技术专业教学指导委员会的研究报告来分,学校培养的人才类型可以分为科学型、工程型和应用型三类,其中科学型人才重在基础理论、技术和方法等的创新;工程型人才以开发实现预定功能要求的系统为主要目标;应用型人才以系统集成为主要途径实现特定功能的需求。

虽然这些不同类型人才的培养有许多共同之处,但是因不同类型人才的就业岗位所需要的责任意识、专业知识能力与素质、人文素养、治学态度、国际化程度等方面存在一定的差异,因而培养目标、培养模式等方面也存在不同。对大多数高校来说,很难兼顾各类人才的培养。因此,合理定位培养目标是确保教学目标和人才培养质量的关键。

由于当前社会领域从事工程开发和应用的岗位数量远远超过从事科学人才的数量,结合当前绝大多数高校的办学现状,安徽省高等学校计算机教育研究会在和多所高校专业负责人以及来自企业的专家反复研究和论证的基础上,形成了以培养工程应用型人才为主的安徽省高等学校计算机类专业的培养目标,并组织研讨组共同探索相关问题,共同建设相关教学资源,共享研究和建设成果,为全面推动安徽省高等学校计算机教育教学水平做出积极的贡献。北京师范大学出版集团安徽大学出版社积极支持安徽省高等学校计算机教育研究会的工作,成立了编委会,组织策划并出版了该套工程应用型计算机系列教材。

为了做好教材的出版工作,编委会在许多方面都采取了积极的措施:

编委会组成的多元化:编委会不仅有来自高校的教育领域的资深教师和专家,而且还有从事工程开发、应用技术的资深专家,从而为教材内容的重组提供更为有力的支持。

教学资源建设的针对性:教材以及教学资源建设的目标就是要突出体现"学以致用"的原则,减少"学不好,用不上"的空泛内容,增加其应用案例,尤其是增设涵盖更多知识点和应用能力的系统性、综合性的案例,以培养学生系统解决问题的能力,进而激发其学习兴趣。

建设过程的规范性:编委会对整体的框架建设、对每本教材和资源的建设都采取汇报、交流和研讨的方式,以听取多方意见和建议;每本书的编写组也都进行反复的讨论和修订,努力提高教材和教学资源的质量。

如果我们的工作能对安徽省高等学校计算机类专业人才的培养做出贡献,那将是我们的荣幸。真诚欢迎有共同志向的高校、企业专家提出宝贵意见和建议,更期待你们参与我们的工作。

胡学钢

2015 年 6 月 10 日于合肥

编委会名单

主　任　　胡学钢（合肥工业大学）

委　员　　（以姓氏笔画为序）
　　　　　王　浩（合肥工业大学）
　　　　　王一宾（安庆师范学院）
　　　　　叶明全（皖南医学院）
　　　　　孙　力（安徽农业大学）
　　　　　刘仁金（皖西学院）
　　　　　朱昌杰（淮北师范大学）
　　　　　沈　杰（合肥炜煌电子有限公司）
　　　　　李　鸿（宿州学院）
　　　　　陈　磊（淮南师范学院）
　　　　　陈桂林（滁州学院）
　　　　　张先宜（合肥工业大学）
　　　　　张润梅（安徽建筑大学）
　　　　　张燕平（安徽大学）
　　　　　金庆江（合肥文康科技有限公司）
　　　　　周国祥（合肥工业大学）
　　　　　周鸣争（安徽工程大学）
　　　　　宗　瑜（皖西学院）
　　　　　郑尚志（巢湖学院）
　　　　　钟志水（铜陵学院）
　　　　　姚志峰（蓝盾信息安全技术股份有限公司）
　　　　　郭有强（蚌埠学院）
　　　　　黄　勇（安徽科技学院）
　　　　　黄海生（池州学院）
　　　　　潘地林（安徽理工大学）

前 言

本书是为高等学校理工科本（专）科的 51 系列单片机原理及接口技术的实践部分教学编写的。全书内容共分 4 部分：第 1 部分介绍基础试验平台；第 2 部分根据教学知识点设计基础实验，包含流水灯控制、中断系统、定时/计数器、串口通信、输入接口识别、矩阵键盘检测、输出接口及 A/D 转换、D/A 转换等应用实验；第 3 部分分别提供了一些基于 STAR ES598PCI 实验仪和 Proteus 软件仿真平台的综合设计实验，Proteus 仿真实验是为了方便同学在没有单片机实验仪情况下进行单片机电路仿真、PCB 设计和虚拟模型仿真；第 4 部分为拓展实验，提供了一些不同芯片在硬件平台上的拓展应用开发实验，供各种创新兴趣小组学习参考，为以后从事单片机及其他嵌入式开发工作奠定基础。

本书可作为高等学校计算机类、电子类、通信类专业学生、自学者实验教学及课程设计用书，还可作为从事单片机系统及其接口技术的工程技术人员的参考书。

本书第 1 章和第 4 章由刘乐群编写，第 2 章和第 3 章由黄大君编写，全书由刘乐群统稿并审核。

书中如有错误及疏漏之处敬请读者批评指正，并请于作者联系（联系邮箱：dajunhuang@hfnu.edu.cn）。

目 录

第 1 章 课程概述 ·· 1
  1.1 课程性质 ··· 1
  1.2 教学目标与任务 ·· 1
  1.3 教学内容 ··· 2
  1.4 教材简介 ··· 2

第 2 章 实验平台 ·· 3
  2.1 实验仪配置方案 ·· 3
  2.2 实验仪功能 ·· 3

第 3 章 基础实验 ·· 6
  实验 1 流水灯实验 ··· 6
  实验 2 中断实验 ·· 10
  实验 3 串行口实验 ··· 14
  实验 4 定时器/计数器实验 ··· 25
  实验 5 输入接口实验 ·· 30
  实验 6 输出接口实验 ·· 34
  实验 7 综合设计实验(定时器、中断综合实验——电子钟) ·············· 47
  实验 8 PWM 实验 ·· 87
  实验 9 8255 控制交通灯实验 ·· 90
  实验 10 8155 输入、输出、SRAM 实验 ······································· 92
  实验 11 8279 键盘显示实验 ··· 94
  实验 12 并行 DA 实验 ··· 97
  实验 13 并行 AD 实验(数字电压表实验) ···································· 99
  实验 14 红外通信实验 ·· 102
  实验 15 X5045 串行 EEPROM 读写实验 ···································· 104
  实验 16 串行 EEPROM 93C46 实验 ·· 111

第 4 章 拓展实验 ·················· 118

实验 1　简易电子琴实验 ·················· 118
实验 2　LED 16×16 点阵实验 ·················· 121
实验 3　I^2C 总线串行 EEPROM 24C02A 实验 ·················· 127
实验 4　电子钟(PCF8563(I^2C 总线)、128×64 液晶) ·················· 133
实验 5　电子钟（CLOCK） ·················· 139
实验 6　数字式温度计实验(18B20、ZLG7290) ·················· 140
实验 7　步进电机实验 ·················· 145
实验 8　直流电机测速实验 ·················· 150
实验 9　ISD1420 语音模块实验 ·················· 155
实验 10　CAN 通信实验 ·················· 159
实验 11　USB 2.0 通信实验 ·················· 165
实验 12　触摸屏实验（ADS7843、12864C） ·················· 167
实验 13　GPS 定位实验 ·················· 174
实验 14　GPRS 通信实验 ·················· 177
实验 15　非接触式卡实验 ·················· 181

第 1 章　　课程概述

1.1　课程性质

"单片机原理及应用"是计算机、电子、通信类专业教学中的一门重要的专业实践课。本课程的教学目的是帮助学习者进一步掌握和使用单片微控制器,使学生学会掌握单片机技术在工业控制、经济建设和日常生活中的应用和创新,提高学习者的工程应用能力和创新能力。本课程的先修课程有:模拟电子技术、数字电路、C 语言与程序设计、汇编语言、微机原理与接口。

本着加强专业理论技术应用、拓宽专业口径、注重实践性环节、提高素质教育的教学理念,希望通过本书的学习实践,培养学生探索、创新思维和分析解决问题的能力,拓展创新应用能力。

1.2　教学目标与任务

1. 总体目标

使学生掌握 51 单片机的内部结构、工作原理、编程技术等有关基础知识和能力,学会 51 单片机在不同领域里的开发、应用。通过实践训练,加深学生对 51 系列单片机的理论知识的理解,掌握一定单片机及其接口技术的应用,培养学生动手能力和独立解决问题的能力。

2. 具体目标

①了解"单片机原理及应用"这门课程的性质、地位和应用领域,了解其在市场上的应用现状及该学科未来的发展方向。

②掌握单片机内部的结构、组成,理解单片机存储器体系结构。

③掌握单片机指令系统及使用。

④掌握基本的编程技术,学会编程和调试。

⑤掌握中断系统的工作原理及应用。

⑥掌握定时器的工作原理及应用。

⑦掌握单片机串行接口技术的原理及应用。

⑧理解单片机存储器扩展技术,学会设计存储器扩展电路。

⑨掌握外围设备与单片机接口技术的工作原理及应用。

⑩掌握单片机多种应用系统设计与开发。

3. 教学任务

培养微处理器和单片微处理器的基本知识、基本理论和基本技能,通过具体项目实验加深对 51 系列单片机基本原理的理解,掌握 51 单片机的中断系统、指令系统与程序设计、定时计数器、串口通信、输入输出接口、应用系统开发等,培养分析问题、解决问题、综合运用所学知识分析处理工程实际问题的能力,提高工程应用素质、创新素质。

1.3 教学内容

教学内容包括理论教学内容和实践教学内容两部分。

表 1-1 教学内容

理论教学内容
单片机概论
单片机的硬件结构
指令系统与程序设计
中断系统与定时计数器
串行通信口
扩展 I/O 接口设计与扩展存储器设计
键盘、显示器等接口设计
应用系统设计与开发

实践教学内容
流水灯实验
中断实验
定时器/计数器实验
串行口实验
输入接口实验
输出接口实验
综合设计实验

1.4 教材简介

为便于不同专业、不同层次的学习者使用,本书分为硬件平台实验和软件仿真实验两部分,硬件平台实验基于 STAR ES598PCI 实验仪设计,适用于拥有该设备使用者学习使用,仿真实验是基于 Proteus 仿真平台设计的,适用于没有专用硬件平台或其他自学者学习使用。每个实验项目和知识点都配有一个调试好的实验程序,并设计了若干课后练习思考题供学习者课前预习、课后提高练习。此外本书针对不同层次、不同专业的学习者还提供了部分基于 Proteus 仿真平台设计和基于 STAR ES598PCI 实验仪上设计开发的综合设计性试验。

第 2 章 实验平台

本书所使用的硬件平台是 STAR ES598PCI 实验仪。每个知识模块提供若干实验项目和实验练习题,满足各高等院校进行单片机课程的开放式实验教学;也可以让参加电子竞赛的学生熟悉各种类型的接口芯片,做各种实时控制实验,轻松面对电子竞赛,完成毕业设计;还可以让刚参加工作的电子工程师迅速成为高手。STAR ES598PCI 还提供实验仪与微机同步演示功能,方便实验室教师的教学、演示。

2.1 实验仪配置方案

STAR ES598PCI 实验仪有 3 种配置方案:
①实验仪主机、仿真模块(不含逻辑分析仪功能、实时跟踪仪功能)。(较低配置)
②实验仪主机、仿真模块(带有逻辑分析仪功能、实时跟踪仪功能)。(中等配置)
逻辑分析仪功能:通过观察采样到的波形,可以让学生了解 CPU 执行指令的完整过程,加深对波形图的认识。
实时跟踪仪功能:记录程序运行轨迹。
③实验仪主机带有自动下载功能,可以另外配置各种仿真器。(高配、使用灵活、适合电子竞赛)

2.2 实验仪功能

2.2.1 软件系统

①完全支持 Keil,支持在 uVision2、uVision3 中使用实验仪。
②提供星研集成环境软件,2004 年它已被认定为高新技术成果转化项目。
• 集编辑器、项目管理、启动编译、连接、错误定位、下载、调试于一体,多种实验仪、仿真器、多类型 CPU 仿真全部集成在一个环境下,操作方法完全一样。
• 完全 VC++ 风格。支持 C、PL/M、宏汇编,同时支持 Keil 公司的 C51、Franklin 公司的 C51、IAR/Archimedes 公司的 C51、Intel C96、Tasking 的 C196、Borland 的 Turbo C。
• 支持 ASM(汇编)、C、PLM 语言,多种语言多模块混合调试,文件长度无限制。
• 支持 BIN、HEX、OMF、AUBROF 等文件格式。可以直接转载 ABS、OMF 文件。
• 支持所有数据类型观察和修改。自动收集变量于变量窗(自动、局部、模块、全局)。
• 无须点击的感应式鼠标提示功能。
• 功能强大的项目管理功能,含有调试该项目有关的仿真器或仿真模块、所有相关文件、编译软件、编译连接控制项等所有的硬软件信息,下次打开该项目,无须设置,即可调试。
• 支持 USB、并口、串口通信。

- 提供模拟调试器。
- 符合编程语言语法的彩色文本显示,所有窗口的字体、大小、颜色可以随意设置。

③提供五十多种实验的汇编、C51 版本的源文件。提供一个库文件,如果学生上机时间有限,只须编写最主要的程序,其他调用库文件即可。

实验仪可提供以下软件实验:十进制数加法、十进制数减法、双字节 BCD 码乘法、双字节二进制数转十进制数、数据传送、冒泡排序、二分查找法、散转、电子钟、频率计等。

2.2.2 硬件系统

1. 传统实验

74HC244、74HC273 扩展简单的 I/O 口;蜂鸣器驱动电路;74HC138 译码;74HC164 串并转换;74HC165 并串转换实验;RS232 和 RS485 接口电路;8155、8255 扩展实验;8253 定时、分频实验;128×64 液晶点阵显示模块;16×16LED 点阵显示模块;键盘 LED 控制器 8279,并配置了 8 位 LED、4×4 键盘;32K 数据 RAM 读写,使用 C51 编制较大实验;并行 AD 实验;并行 DA 实验;直流电机控制;步进电机控制;PWM 脉宽调制输出接口;继电器控制实验;逻辑笔;打印机实验;电子琴实验;通过 74HC4040 分频得到十多种频率信号;提供 8 个拨码盘、8 个发光二极管、8 个独立按键;单脉冲输出。

STAR ES598PCI 特有功能:8250 串行通讯实验;8251 串行通讯实验。

STAR ES598PA 特有功能:主板允许 P0、P2 口作 I/O 口线使用;普通光耦实验、高速光耦实验;8155 键盘 LED 实验(共有三种键盘 LED 控制方式)。

2. 拓展实验

录音、放音模块实验;光敏实验;压力传感器实验;频率计实验;接触式 IC 卡读写实验;非接触式 IC 卡读写实验(扩展模块)、触摸屏实验(扩展模块)、NAND FALSH 实验(扩展模块)。

3. 串行接口实验

①一线:DALLAS 公司的 DS18B20 测温实验。

②I^2C:实时钟 PCF8563、串行 EEPROM 24C02A、键盘 LED 控制器实验。

③SPI:串行 D/A、串行 A/D 实验、串行 EEPROM 及看门狗 X5045。

④Microwire 总线的串行 EEPROM:AT93C46。

⑤红外通信实验。

⑥CAN:CAN 2.0(扩展模块)。

⑦USB:USB 1.1、USB 2.0、USB 主控(扩展模块)。

⑧以太网:10M 以太网模块(扩展模块)。

⑨蓝牙:(扩展模块)。

4. 闭环控制

①门禁系统实验。

②光敏实验或压力传感器实验。

③旋转图形展现实验。

④RTX-51 Real-Time OS。

⑤直流电机转速测量,使用光电开关或霍尔器件测量电机转速。
⑥直流电机转速控制,使用光电开关或霍尔器件精确控制电机转速。
⑦数字式温度控制,通过该实验能够认识控制在实际中的应用。

5. 实验扩展区,提供扩展实验接口,用户可自行设计实验

可以提供 USB1.1、USB2.0、USB 主控、10M 以太网接口的 TCP/IP 实验模块、CAN 总线、非接触式 IC 卡、触摸屏模块、GPS、GPRS、双通道虚拟示波器、虚拟仪器、读写优盘、CPLD、FPGA 模块。

6. EDA-CPLD、FPGA 可编程逻辑实验

逻辑门电路:与门、或门、非门、异或门、锁存器、触发器、缓冲器等;半加器、全加器、比较器、二十进制计数器、分频器、移位寄存器、译码器;常用 74 系列芯片、接口芯片实验;8 段数码块显示实验;16×16 点阵式 LED 显示实验;串行通信收发;I^2C 总线等。

提供汇编语言及 C51 语言编写的实验范例。

7. 单片机引脚说明

JP45:地址线 A0..A7。

JP48、JP50:CPU 的 P0 口,它只能作地址/数据总线使用,不能做 I/O 口使用。

JP51、JP55:CPU 的 P1 口。

JP59:CPU 的 P2,它可作地址线 A8..A15 使用。

JP61、JP64:CPU 的 P3 口,P3.7、P3.6 作读、写信号线用。

JP66:相当于一个 CPU 座,使用 40 芯扁线与用户板相连,可仿真 P0、P2 口作地址/数据使用的 CPU。

第 3 章　基础实验

本章将结合实验仪的所有单元电路(包括标准配置和可选各种模块)向使用者逐一介绍各个实验,由浅入深,从最基础的实验开始,直到学会使用当今流行的各种单片机外围电路,开发有一定深度的单片机项目,硬件实验分为基础实验和综合实验两部分。读者也可以根据自己的理解、需要,将各个单元电路自行组合而成具有实际意义的复杂单片机控制电路,在设计电路板前,在实验仪上作一认证。

实验 1　流水灯实验

一、实验目的

1. 安装及使用星研集成环境软件或熟悉 Keil C51 集成环境软件的使用方法。
2. 熟悉 MCS51 汇编指令,能自己编写简单的程序,控制硬件。

二、实验设备

STAR 系列实验仪一套、PC 机一台。

三、实验内容

1. 熟悉星研集成环境软件或熟悉 Keil C51 集成环境软件。
2. 按照接线图编写程序:使用 P2 口控制 G6 区的 8 个指示灯,循环点亮,瞬间只有一个灯亮。
3. 观察实验结果,验证程序是否正确。

四、实验原理图

流水灯实验原理图见图 3-1。

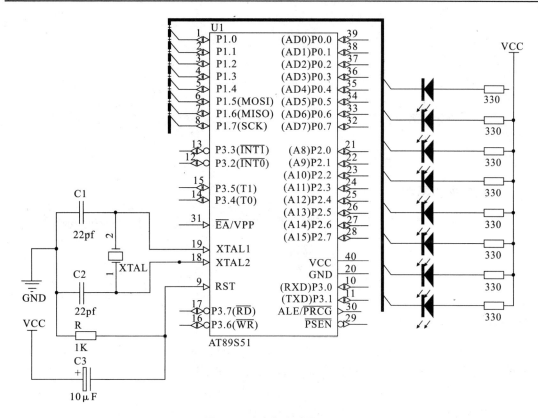

图 3-1 流水灯实验原理图

五、实验步骤

1. 连线说明：

| A3 区:JP51(P1 口) | —— | G6 区:JP65(发光管) |

2. 编写程序或运行参考程序。

3. 实验结果：通过 G6 区的 LED 指示灯(8 个指示灯轮流点亮)，观察实验的输出结果是否正确。

六、演示程序

```
#include "reg52.h"
#include "intrins.h"
void delay()            //延时
{
    unsigned int i;
    for (i=0; i < 0xffff; i++)
    {;}
}

main()
```

```
    {
        P1=0xfe;
        while(1)
        {
            P1=_crol_(P1,1);
            delay();
        }
    }
```

七、实验扩展及思考

Delay 是一个延时子程序,改变延时常数,使用全速运行命令,显示发生的变化。

流水灯的仿真实验

一、实验目的

按照下图连接单片机电路图,P3 口连接 8 个 LED 发光二极管,设计一个 8 位二进制加法器,完成从 000000000 到 11111111 的计数,试编写程序让 8 个 LED 灯实现加法器显示。要求:

1. 灯亮表示 1,灯灭表示 0,实现从 00000000～11111111 的加法过程。
2. 开始计数前全灭全亮闪烁两次,结束计数时闪烁两次全灭。
3. 要有延时子程序、闪烁子程序。

二、实验原理图

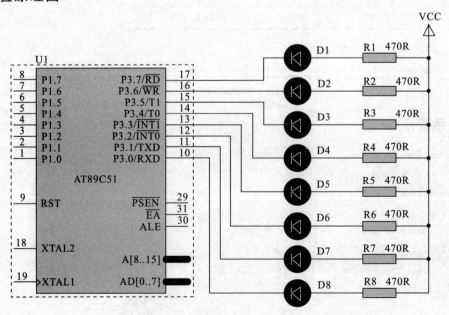

图 3-2　流水灯仿真实验原理图

三、参考实验程序

```c
#include<reg51.h>    //包含单片机寄存器的头文件
/******************功能:延时函数*********************/
void delay(void)
{
    unsigned int i,j;
    for(i=0;i<20000;i++)
        for(j=0;i<200;j++)
        { ;}
}
void flash(void)
{
    P0=0xff;
    delay();
    P0=0x00;
    delay();
    P0=0xff;
    delay();
    P0=0x00;
    delay();
    delay();
}
/********************主函数*********************/
void main(void)
{
    flash();
    unsigned char i;
    while(i<=0xff)       //注意 i 的值不能超过 255
    {
        P0=i;            //将 i 的值送 P0 口
        i++;
        delay();         //调用延时函数
    }
    flash();
}
```

四、实验思考题

1. 修改程序使 LED 灯为 D1、D8 灯亮——D2、D7 灯亮——D3、D6 灯亮——D4、D5 灯亮,再将 LED 灯亮的顺序倒过来,即 D4、D5 灯亮——D3、D6 灯亮——D2、D7 灯亮——D1、D8 灯亮,连续运行。

2. 自行设计一个节日彩灯。

实验 2　中断实验

一、实验目的

1. 熟悉定时器/计数器的定时功能。
2. 熟悉编写简单定时器中断程序,控制硬件。

二、实验设备

STAR 系列实验仪一套、PC 机一台。

三、实验内容

1. 熟悉星研集成环境软件或熟悉 Keil C51 集成环境软件的安装和使用方法。
2. 按照接线图编写程序:使用 P0 口控制 G6 区的 8 个指示灯,依次亮 1 秒灭 1 秒并一直如此显示,请用定时器 T0 的中断函数来编写。
3. 观察实验结果,验证程序是否正确。

四、实验原理图

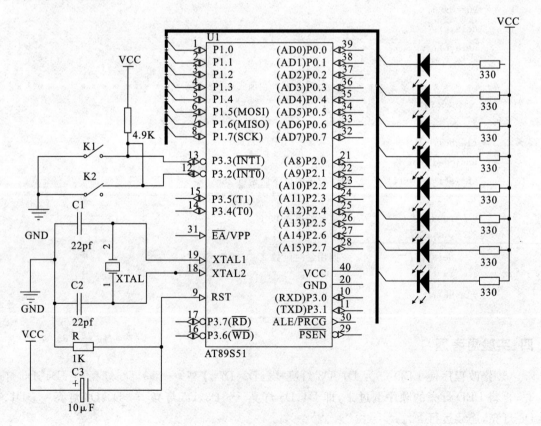

图 3-3　中断原理图

五、实验步骤

1. 连线说明：

A3 区:JP51(P1 口)——G6 区:JP65(流水灯)

A3 区:JP61(P3 口)——G6 区:JP80(开关)

2. 编写程序或运行参考程序。

3. 实验结果:通过 G6 区的 LED 指示灯(亮 1 秒灭 1 秒),观察实验的输出结果是否正确。

六、实验程序

```
#include "reg52.h"
unsigned char i=20;
main()
{
    TMOD=0x01;
    TH0=(65536-50000)/256;
    TH0=(65536-50000)%256;
    P2=0xff;
    EA=1;
    ET0=1;
    TR0=1;
    while(1);
}
void T0_INT(void) interrupt 1
{
    TH0=(65536-50000)/256;
    TH0=(65536-50000)%256;
    i--;
    if (i==0)
    {
        P1=~P1;
        i=20;
    }
}
```

七、实验扩展及思考

1. 修改程序,使 LED 灯亮灭 0.5 秒、2 秒或其他的时间间隔。

2. 利用 P3.1、P3.3 脚设计一个能通过 IP 寄存器、IE 寄存器控制优先级和中断开关的中断控制系统,通过中断处理程序以不同优先级和不同闪烁方式显示中断处理结果。

中断的仿真实验

一、实验目的

通过对 P3.2、P3.3 引脚的电平控制,实现外部中断处理,从而控制输出口 P1 的输出效果变化。

二、实验参考原理图

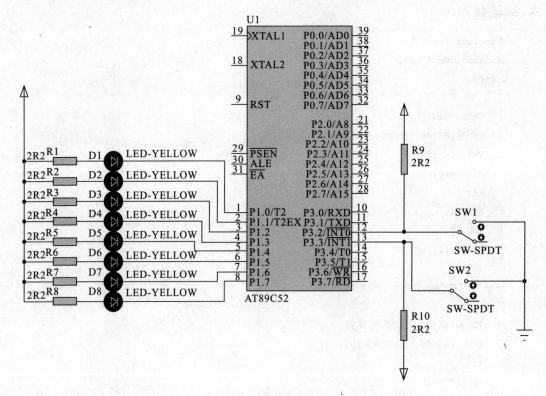

图 3-4 中断仿真原理图

三、参考实验程序

```
/*用外中断0的中断方式进行数据采集和处理*/
#include<reg52.h>
#include<intrins.h>
void init();
void delay(unsigned int);

void main()
{
    init();
    while(1)
```

```
    {
        P1=0xff;
        delay(100);
        P1=0x00;
        delay(100);
    }
}
void init()
{
    EA=1;
    IT0=0;
    IT1=0;
    EX0=1;
    EX1=1;
}
void delay(unsigned int n)
{
    unsigned int i,j;
    for(i=0;i<n;i++)
        for(j=0;j<110;j++);
}
void aa() interrupt 0
{
    unsigned char tmp=0xfe;
    unsigned int i=0;
    P1=tmp;
    delay(100);
    i=7;
    while(i--)
    {
        tmp=_crol_(tmp,1);
        P1=tmp;
        delay(100);
    }
    //delay(500);
    i=7;
    while(i--)
    {
        tmp=_cror_(tmp,1);
        P1=tmp;
        delay(100);
    }
//delay(500);
```

```
    }
    void bb() interrupt 2
    {
        P1=0xf0;
        delay(500);
        P1=0x0f;
        delay(500);
    }
```

四、实验思考题

根据指导书中提供的原理图,自行设计一个外部中断实验,要求:
(1)两个外部中断全部用上。
(2)实验能体现不同中断优先级的中断源的响应情况。
(3)不同中断处理程序能输出不同的响应效果。

实验 3　串行口实验

一、实验目的

掌握单片机串行通讯;初步了解远程控制方法。

二、实验设备

STAR 系列实验仪二套、PC 机二台。

三、实验内容

1. 单片机串行口通信参数
(1)传输距离≤0.5m,最大传输率≤100Kbps。
(2)半双工工作方式。

2. 实验过程
(1)主机通过单片机串行口发出控制命令给从机。
(2)从机收到控制命令,检验命令的正确性,执行命令:点亮相应的发光管。

四、实验原理图

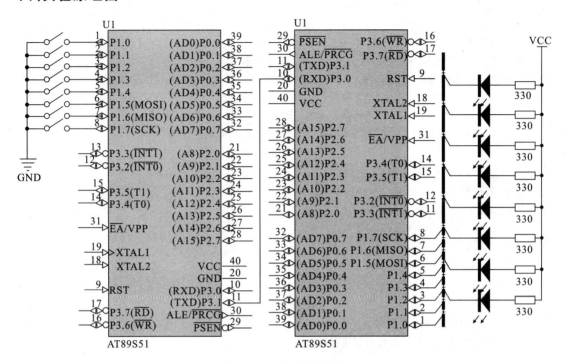

图 3-5　串口通信原理图

五、实验步骤

1. 主机连线说明：

G6 区:JP80(开关)——A3 区:JP51(P1 口)

主机:TXD——从机:RXD

从机连线说明：

G6 区:JP65(发光管)——A3 区:JP51(P1 口)

2. 命令正确,点亮相应发光管。

六、演示程序

发送主机程序：

```
#include <reg51.H>
#define uchar unsigned char
#define uint unsigned int
void main()
{
    uchar temp;
    TMOD=0X20;          //设置定时器1为工作方式2
    SCON=0X40;
    TH1=0XFD;           //装初值
```

```c
    TL1=0XFD;
    TR1=1;              //启动 T
    P1=0XFF;
    while(1)
    {
        temp=P1;
        SBUF=temp;      //发送数据
        while(!TI);     //检测数据是否发送完毕,未完在这个地方等待
        TI=0;           //发送到结束位时 TI 自动置 1,需软件置 0
    }
}
```

接收主机程序:

```c
#include <reg51.H>
#define uchar unsigned char
#define uint unsigned int
uchar a;
void main()
{
    uchar temp=0;
    TMOD=0X20;          //设置定时器 1 为工作方式 2
    SCON=0X50;
    TH1=0XFD;           //装初值
    TL1=0XFD;
    TR1=1;              //启动 T
    while(1)
    {
        RI=0;           //接收开始位时 RI 自动置 1,需软件置 0
        a=SBUF;         //接收数据
        P1=a;
    }
}
```

七、实验扩展及思考

1. 单片机串行口通信如何实现既接收又发送?
2. 在掌握单片机串行通讯和基本的远程控制方法的基础上,进行其他方面的远程控制,如远程控制电机转速、语音、温度测量、显示等。

串行口仿真实验

一、实验目的

本实验要求单片机 U1 通过其串行口 TXD 向计算机发送一个数据"0xab"。利用串行

口将单片机的输出信号转化成计算机能够识别的信号。针对发送的实例,再设计一个单片机接受计算机送出数据的过程。

二、实验原理图

单片机发送数据实验参考电路图如图 3-6 所示。

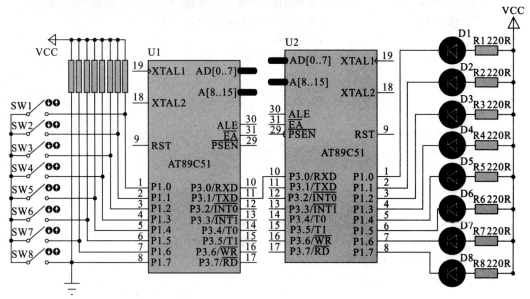

图 3-6 串口通信仿真原理图

三、实验参考程序

```
/**********发送单片机的发送程序************/
#include<reg51.h>       //包含寄存器的头文件
/**********向PC发送一个字节数据************/
void Send(unsigned char date)
{
    SBUF=date;
    while(TI==0)
        ;
    TI=0;
}
/********延时约150ms********/
void delay(void)
{
    unsigned char m,n;
    for(m=0;m<200;m++)
        for(n=0;n<250;n++)
            ;
}
/********函数功能:主函数********/
void main(void)
```

```c
{
    unsigned char temp;
    TMOD=0x20;              //定时器T1工作于方式2
    SCON=0x40;              //串口工作方式1
    PCON=0x00;              //波特率9600
    TH1=0xfd;               //根据规定给定时器T1赋初值
    TL1=0xfd;               //根据规定给定时器T1赋初值
    TR1=1;                  //启动定时器T1
    P1=0xff;                //读取P1端口数据
    while(1)
    {
        temp=P1;
        Send(temp);         //发送数据i
        delay();            //50ms发送一次检测数据
    }
}

/************接收单片机的接收程序************/
#include<reg51.h>          //包含单片机寄存器的头文件
/******接收一个字节数据******/
unsigned char Receive(void)
{
    unsigned char date;
    while(RI==0)            //只要接收中断标志位RI没有被置"1"
        ;                   //等待,直至接收完毕(RI=1)
    RI=0;                   //为了接收下一帧数据,需将RI清0
    date=SBUF;              //将接收缓冲器中的数据存于date
    return date;
}
/******主函数******/
void main(void)
{
    TMOD=0x20;              //定时器T1工作于方式2
    SCON=0x50;              //SCON=0101 0000B,串口工作方式1,REN=1
    PCON=0x00;              //PCON=0000 0000B,波特率9600
    TH1=0xfd;               //根据规定给定时器T1赋初值
    TL1=0xfd;               //根据规定给定时器T1赋初值
    TR1=1;                  //启动定时器T1
    REN=1;                  //允许接收
    while(1)
    {
        P1=Receive();       //将接收到的数据送P1口显示
    }
}
```

四、课后思考题

设计一个多机通信过程,要求一个机器发送 3 种不同规律的数组到 3 个单片机上,并通过接收单片机上的流水灯展示来检测接收结果。参考原理图如图 3-7 所示。

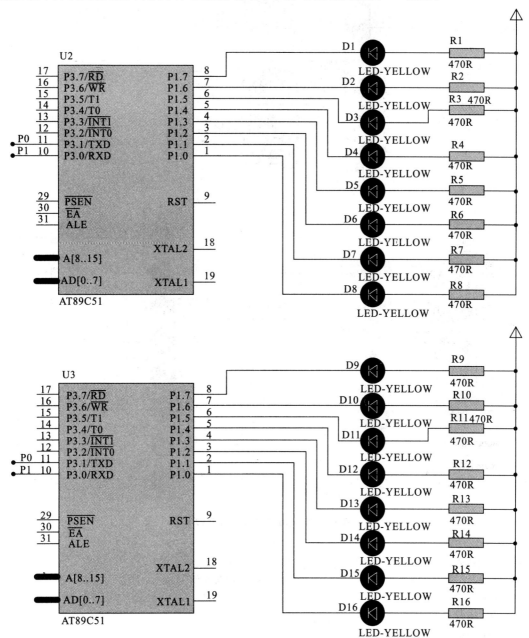

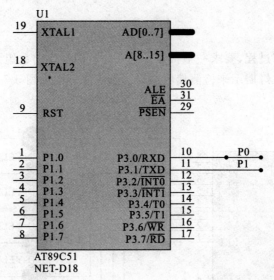

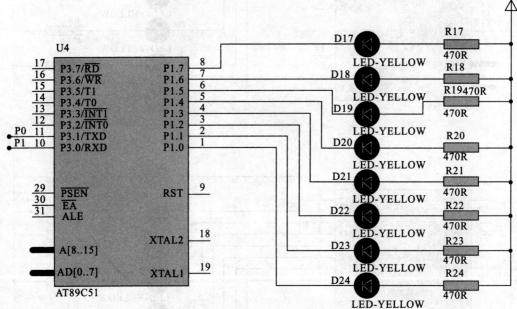

图 3-7　多机串口通信仿真原理图

参考程序：

//多机通信:数据发送程序(波特率9600,串口方式3,T1方式2)
```
#include<reg51.h>         //包含单片机寄存器的头文件
sbit p=PSW^0;
unsigned int i=0;
unsigned int counts=0;
unsigned int adds=0;
unsigned char a=0xff;
unsigned int code Tab[]={0xFE,0xFD,0xFB,0xF7,0xEF,0xDF,0xBF,0x7F};
unsigned int code Tab1[]={0x7F,0xBF,0xDF,0xEF,0xF7,0xFB,0xFD,0xFE};
void delay()
```

```c
{
    unsigned int i, j;
    for(i=0; i<50; i++)
        for(j=0; j<250; j++);
}
void main()
{
    TMOD=0x20;              //TMOD=0010 0000B,定时器 T1 工作于方式 2
    TH1=0xfd;               //根据规定给定时器 T1 赋初值
    TL1=0xfd;               //根据规定给定时器 T1 赋初值
    TR1=1;                  //启动定时器 T1
    SCON=0xe0;              //SCON=1110 0000B,串口工作方式 3,
                            //SM2 置 0,不使用多机通信,TB8 置 0
    PCON=0x00;              //PCON=0000 0000B,波特率 9600
    IE=0x90;
    TB8=1;                  //寻址
    SBUF=0x00;              //第一次随便发,以便产生发送中断
    while(1);

}
void Time() interrupt 4     //"interrupt"声明函数为中断服务函数
{
    TI=0;                   //清中断标志
    counts++;
    if(counts>=32)          //给每个从机 发送信息
    {
        adds++;             //满条 换下一个从机
        if(adds==3){
            adds=0;
            }
        counts=0;
    }

    if(adds==0)             //表示从机号
    {
        if(counts==0)
        {
            TB8=1;          //表示发送的是地址信息
            SBUF=0x00;      //第一次发送给 0 号从机
            delay();
        } else TB8=0;       //表示发送的是数据
        SBUF=Tab[i];
        i++;
```

```c
            delay();
            if(i==8)
            i=0;
    }
    else if(adds==1)
    {
        if(counts==0)
        {
            TB8=1;              //表示发送的是地址信息
            SBUF=0x01;          //第一次发送给1号从机
            delay();
        } else TB8=0;           //表示发送的是数据

SBUF=Tab1[i];
i++;
delay();
if(i==8)
i=0;
}
else if(adds==2)
{
    if(counts==0)
    {
        TB8=1;                  //表示发送的是地址信息
        SBUF=0x02;              //第一次发送给2号从机
        delay();
    } else TB8=0;               //表示发送的是数据
    SBUF=a;
    delay();
    a=~a;
    }
}
//多机通信:1号单片机接收程序(波特率9600,串口方式3,T1方式2)

#include<reg51.h>
sbit p=PSW^0;
unsigned char Receive(void)
{
    unsigned char dat;
    while(RI==0);
    RI=0;                       //为了接收下一帧数据,需将RI清0
    dat=SBUF;                   //将接收缓冲器中的数据存于dat
    if(SM2)
    {
```

```c
        if(dat==0x00)
        {
            SM2=0;
            return 0x00;
        }
        else
        {   return 0xff;
        }
    }
    else{return dat;}
}
void main(void)
{
    unsigned int x=0;
    TMOD=0x20;              //定时器T1工作于方式2
    SCON=0xf0;              //SCON=1111 0000B,串口工作方式3,允许接收(REN=1)
    PCON=0x00;              //PCON=0000 0000B,波特率9600
    TH1=0xfd;               //根据规定给定时器T1赋初值
    TL1=0xfd;               //根据规定给定时器T1赋初值
    TR1=1;                  //启动定时器T1
    REN=1;                  //允许接收
    while(1)
    {
        P1=Receive();       //将接收到的数据送P1口显示
        x++;
        if(x>32)
        {   x=0;
            SM2=1;
        }
    }
}
//多机通信:2号单片机接收程序(波特率9600,串口方式3,T1方式2)
#include<reg51.h>
sbit p=PSW^0;
unsigned char Receive(void)
{
    unsigned char dat;
    while(RI==0)            //只要接收中断标志位RI没有被置1,
        ;                   //等待,直至接收完毕(RI=1)
    RI=0;                   //为了接收下一帧数据,需将RI清0

    dat=SBUF;               //将接收缓冲器中的数据存于dat
    if(SM2==1)
    {
```

```c
                if(dat==0x01)
                {
                    SM2=0;
                    return 0x00;
                }
                else
                {
                    return 0xff;
                }
            }
            else
            {
                return dat;
            }
}

void main(void)
{   unsigned int dt=0;
    TMOD=0x20;          //定时器T1工作于方式2
    SCON=0xf0;
    PCON=0x00;          //PCON=0000 0000B,波特率9600
    TH1=0xfd;           //根据规定给定时器T1赋初值
    TL1=0xfd;           //根据规定给定时器T1赋初值
    TR1=1;              //启动定时器T1
    REN=1;              //允许接收
    while(1)
    {
        P1=Receive();   //将接收到的数据送P1口显示
        dt++;
        if(dt>32)
        {
            dt=0;
            SM2=1;
        }
    }
}

//多机通信:3号单片机接收程序(波特率9600,串口方式3,T1方式2)
#include<reg51.h>
sbit p=PSW^0;
unsigned char Receive(void)
{
    unsigned char dat;
    while(RI==0);       //等待,直至接收完毕(RI=1)
```

```
            RI=0;                    //为了接收下一帧数据,需将 RI 清 0
            dat=SBUF;                //将接收缓冲器中的数据存于 dat
            if(SM2)
            {   if(dat==0x02)
                {   SM2=0;
                    return 0x00;
                }
                else
                {   return 0xff;
                }
            }
            else
            {
                return dat;
            }
        }

        void main(void)
        {   unsigned int x=0;
            TMOD=0x20;               //定时器 T1 工作于方式 2
            SCON=0xf0;               //SCON=1101 0000B,串口工作方式 3,允许接收(REN=1)
            PCON=0x00;               //PCON=0000 0000B,波特率 9600
            TH1=0xfd;                //根据规定给定时器 T1 赋初值
            TL1=0xfd;                //根据规定给定时器 T1 赋初值
            TR1=1;                   //启动定时器 T1
            REN=1;                   //允许接收
            while(1)
            {
                P1=Receive();        //将接收到的数据送 P1 口显示
                x++;
                if(x>32)
                {
                    x=0;
                    SM2=1;
                }
            }
        }
```

实验 4　定时器/计数器实验

一、实验目的

1. 熟悉定时器/计数器的定时功能。
2. 熟悉编写简单定时器中断程序,控制硬件。

二、实验设备

STAR 系列实验仪一套、PC 机一台。

三、实验内容

1. 熟悉星研集成环境软件或熟悉 Keil C51 集成环境软件的安装和使用方法。
2. 按照接线图编写程序：使用 P1 口控制 G6 区的 8 个指示灯，依次亮 1 秒灭 1 秒并一直如此显示，请用定时器 T0 的中断函数来编写。
3. 观察实验结果，验证程序是否正确。

四、实验原理图

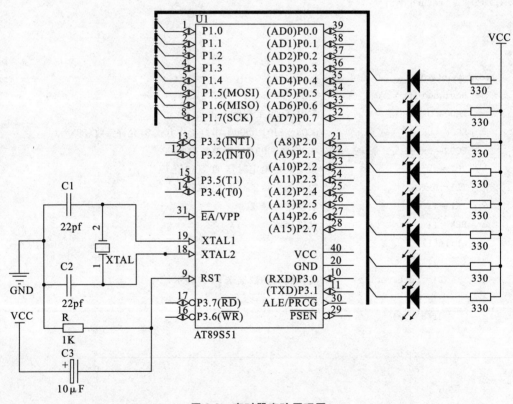

图 3-8　定时器实验原理图

五、实验步骤

1. 连线说明：

A3 区:JP51(P1 口)——G6 区:JP65(发光管)

2. 编写程序或运行参考程序。

3. 实验结果：通过 G6 区的 LED 指示灯(亮 1 秒灭 1 秒)，观察实验的输出结果是否正确。

六、实验程序

```c
#include "reg52.h"
unsigned char i=20;
main()
{
    TMOD=0x01;
    TH0=(65536-50000)/256;
    TH0=(65536-50000)%256;
    P1=0xff;
    EA=1;
    ET0=1;
    TR0=1;
    while(1)
    {
        ;
    }
}
void T0_INT(void) interrupt 1
{
    TH0=(65536-50000)/256;
    TH0=(65536-50000)%256;
    i--;
    if (i==0)
    {
        P1=~P1;
        i=20;
    }
}
```

七、实验扩展及思考

1. 修改程序,使LED灯亮灭0.5秒、2秒或其他的时间间隔。
2. 改变连线和程序,设计一个1kHz的连续的音频信号,并用蜂鸣器输出。

定时器/计数器仿真实验

一、实验目的

用定时器T0的中断实现定时,系统时钟为12MHz,通过使用定时器T0的中断来控制P2.0引脚的LED的灯闪烁,要求闪烁时间2秒,即亮1秒,灭1秒。

二、实验原理图

实验参考电路图如图 3-9 所示(下图只有 D1 灯闪烁,即亮 1 秒,灭 1 秒)。

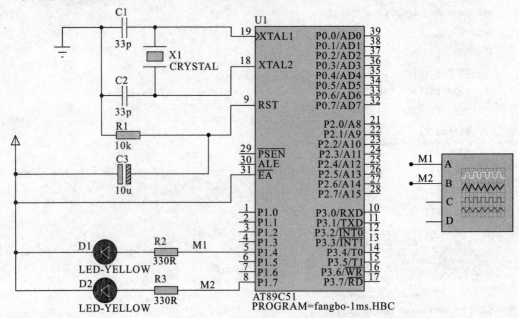

图 3-9 定时器仿真实验原理图

三、参考实验程序

```
#include<reg51.h>
sbit D1=P2^0;                    //将 D1 位定义为 P2.0 引脚
unsigned char Countor;           //定义一个全局变量 Countor 记录 T0 中断次数
void main(void)                  //主函数
{
    EA=1;                        //开中断
    ET0=1;
    TMOD=0x01;                   //使用 T0 工作定时模式,方式 2
    TH0=(65536-15536)/256;       //T0 赋初值
    TL0=(65536-15536)%256;
    TR0=1;                       //启动 T0
    Countor=0;                   //中断次数记录
    while(1);
}
void Time0(void) interrupt 1 using 0    //T0 的中断服务函数
{
    Countor++;                   //中断次数累加
    if(Countor==20)              //若累计满 20 次,即计时满 1s
    {
```

```
            D1=~D1;                    //将 P2.0 输出取反
            Countor=0;                 //将 Countor 清 0,重新计数
    }
        TH0=(65536-15536)/256;         //T0 重新赋初值
        TL0=(65536-15536)%256;
}
```

四、实验思考题

1. 修改程序使用定时器 T1 的中断方式来控制 P2.0、P2.1 引脚的 LED 灯分别以 200ms 和 800ms 的周期闪烁。

2. 设计一个 1kHz 断续的音频信号,并用蜂鸣器输出。

参考原理图:

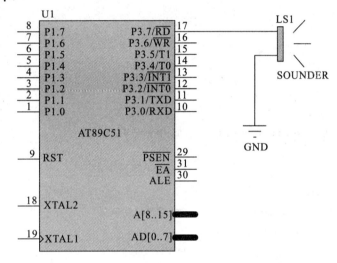

图 3-10 定时器控制蜂鸣器仿真实验原理图

参考程序

```
#include<reg51.h>
#define uint unsigned int
sbit sound=P1^0;              //将 sound 位定义为 P1.0 引脚
uint count;
void delay(uint);
void main(void)
{
    TMOD=0x01;                //T0 工作于方式 1
    EA=1;                     //开总中断
    count=0;
    TH0=(65536-1000)/256;     //T0 赋初值
    TL0=(65536-1000)%256;
    TR0=1;
    ET0=1;
```

```
    while(1)
    {
        while(TF0==0);              //无限循环等待中断
        if(count==200)
        {   TR0=0;
            delay(200);
            TR0=1;
            count=0;
        }
    }
}

void delay(uint z)
{
    uint x,y;
    for(x=z;x>0;x--)
    for(y=124;y>0;y--);
}
void Time0(void) interrupt 1 using 0    //T1中断服务程序
{
    count++;
    sound=~sound;
    TH0=(65536-1000)/256;               //T0重新赋初值
    TL0=(65536-1000)%256;
}
```

实验 5 输入接口实验

一、实验目的

1. 熟悉定时器/计数器的定时功能。
2. 熟悉编写简单定时器中断程序,控制硬件。

二、实验设备

STAR 系列实验仪一套、PC 机一台。

三、实验内容

1. 熟悉星研集成环境软件或熟悉 Keil C51 集成环境软件的安装和使用方法。
2. 通过检测输入开关来控制 LED 灯的亮灭。
3. 观察实验结果,验证程序是否正确。

四、实验原理图

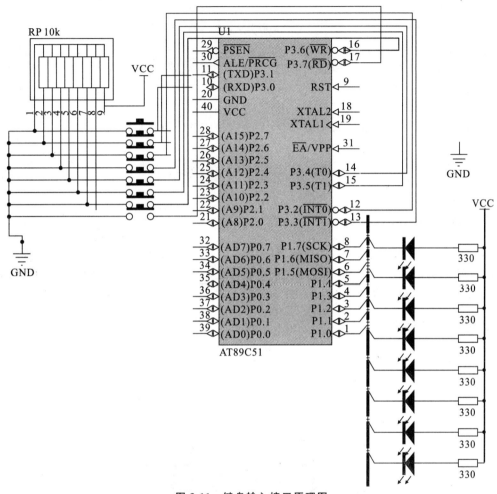

图 3-11 键盘输入接口原理图

五、实验步骤

1. 连线说明：

A3 区:JP51(P1 口)——G6 区:JP65(发光管)

A3 区:JP61(P3 口)——G6 区:JP80(开关)

2. 编写程序或运行参考程序。

3. 实验结果:通过检测按钮开关的输入状态来控制 LED 灯的亮灭观察实验的输出结果是否正确。

六、实验程序

```
#include <reg51.h>
void main( )            /*主函数*/
{   while(1)
```

```
    { unsigned char temp;        /*定义临时变量 temp*/
      P1=0xff;                   /* P1 口位置 1,作为输入*/
      temp=P1;                   /*读 P1 口,送临时变量 temp*/
      P2=temp;                   /*临时变量值写入 P2 口输出*/
    }
}
```

七、实验扩展及思考

利用图中所示的 8 个独立式键盘,设计一个简易的 8 位抢答器。

输入接口仿真实验

一、实验目的

设计一个 4×4 的矩阵键盘,键盘的号码 0~15,要求编写出一个键盘输入扫描程序,要求单片机能根据键盘排列顺序,将按下去的键盘号正确识别出来,并采用两个数码管分别作为键盘号码的个位和十位。

二、实验参考原理图

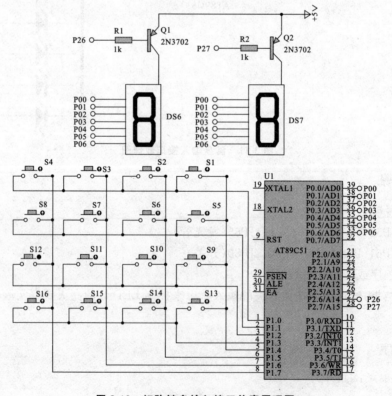

图 3-12 矩阵键盘输入接口仿真原理图

三、参考实验程序

```c
#include<reg51.h>
unsigned char code Tab[ ]={0xc0,0xf9,0xa4,0xb0,0x99,0x92,0x82,0xf8,0x80,0x90};
//数字0~9的段码
unsigned char keyNo;
void led_DelayMS(unsigned char i)                  //扫描延时函数
{
    unsigned char j;
    while(i--)
    {
        for(j=0;j<120;j++)
        ;}
}
void display(unsigned char k)                      //数码管显示子程序
{
    P2=0xbf;
    P0=Tab[k/10];
    led_delay();
    P2=0x7f;
    P0=Tab[k%10];
    led_delay();
}

void main(void)
{
    EA=1;
    ET0=1;
    TMOD=0x01;
    TH0=(65536-500)/256;
    TL0=(65536-500)%256;
    TR0=1;
    keyNo=0x00;

    while(1)
    {
        display(keyval);
    }
}
void time0_interserve(void) interrupt 1 using 1    //T0的中断服务程序完成键盘扫描
{
    unsignedchar Tmp;
    P1=0x0f;                                       //高4位置0,放入4行
```

```
        DelayMS(1);
        switch(Tmp)                //判断按键发生于0~3列的哪一列
        {
            case 1:KeyNo=0;break;
            case 2:KeyNo=1;break;
            case 4: KeyNo=2;break;
            case 8:KeyNo=3;break;
            default:KeyNo=16;      //无键按下
        }
        P1=0xf0;                   //低4位置0,放入4列
        DelayMS(1);
        Tmp=P1>>4^0x0f;            //按键后f0变成XXXX0000,X中有1个为0,3个仍为1
        //高4位转移到低4位并异或得到改变的值
        switch(Tmp)//对0~3行分别附加起始值0,4,8,12
        {
            case 1:KeyNo+=0;break;
            case 2:KeyNo+=4;break;
            case 4:KeyNo+=8;break;
            case 8:KeyNo+=12;
        }
        keyNo=16;
        TR0=1;
        TH0=(65536-500)/256;
        TL0=(65536-500)%256;
}
```

四、实验思考题

修改实验电路图和实验程序和设计电路,改成静态显示。

实验6 输出接口实验

一、实验目的

了解图形液晶模块的控制方法;了解它与单片机的接口逻辑;掌握使用图形点阵液晶显示字体和图形。

二、实验设备

STAR系列实验仪一套、PC机一台。

三、实验内容

1. 12864J 液晶显示器

①图形点阵液晶显示器,分辨率为128×64,可显示图形和8×4个(16×16点阵)汉字。

②采用8位数据总线并行输入输出和8条控制线。
③指令简单,7种指令。

2. 实验过程

在12864J液晶上显示一段字,包括汉字和英文:"星研电子"、"STAR ES51PRO"、"欢迎使用",三行字。

四、实验原理图

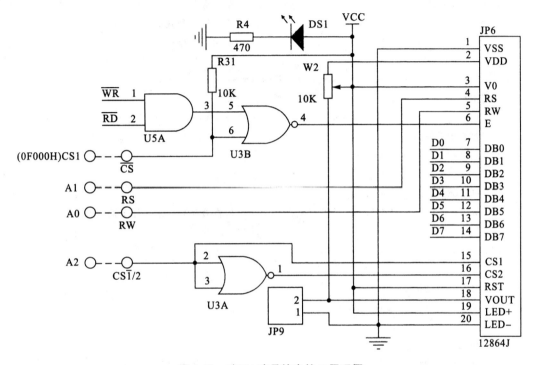

图3-13 12864液晶输出接口原理图

五、实验步骤

1. 主机连线说明:
A1区:CS、RW、RS、CS1/2——A3区:CS1、A0、A1、A2
2. 运行程序,验证显示结果。

六、演示程序

```
//;1.12864J 显示程序:12864j.c
/****************************************************
*    12864J 液晶显示器
*    12864J:1.图形点阵液晶显示器,分辨率为128×64。显示图形和8×4个(16×16点阵)汉字
*    2.采用8位数据总线并行输入输出和8条控制线
*    3.指令简单,7种指令
****************************************************/
```

```c
xdata unsigned char WR_COM_AD_L _at_ 0xF004;        //写左半屏指令地址
xdata unsigned char WR_COM_AD_R _at_ 0xF000;        //写右半屏指令地址
xdata unsigned char WR_DATA_AD_L _at_ 0xF006;       //写左半屏数据地址
xdata unsigned char WR_DATA_AD_R _at_ 0xF002;       //写右半屏数据地址
xdata unsigned char RD_BUSY_AD _at_ 0xF001;         //查忙地址
xdata unsigned char RD_DATA_AD _at_ 0xF003;         //读数据地址

#define X 0xB8                                      //起始显示行基址
#define Y 0x40                                      //起始显示列基址
#define FirstLine 0xC0                              //起始显示行

//*******************************************
//基本控制
//*******************************************
//写左半屏控制指令
void WRComL(unsigned char _data)
{
    WR_COM_AD_L=_data;
    while (RD_BUSY_AD & 0x80)                       //检查液晶显示是否处于忙状态
    {;}
}

//写右半屏控制指令
void WRComR(unsigned char _data)
{
    WR_COM_AD_R=_data;
    while (RD_BUSY_AD & 0x80)                       //检查液晶显示是否处于忙状态
    {;}
}

//写左半屏数据
void WRDataL(unsigned char _data)
{
    WR_DATA_AD_L=_data;
    while (RD_BUSY_AD & 0x80)                       //检查液晶显示是否处于忙状态
    {;}
}

//写右半屏数据
void WRDataR(unsigned char _data)
{
    WR_DATA_AD_R=_data;
    while (RD_BUSY_AD & 0x80)                       //检查液晶显示是否处于忙状态
```

```
        {;};
}

//显示左半屏数据,count-显示数据个数
void DisplayL(unsigned char * pt, char count)
{
    while (count——)
    {
        WRDataL( * pt++);                  //写左半屏数据
    }
}


//显示右半屏数据,count-显示数据个数
void DisplayR(unsigned char * pt, char count)
{
    while (count——)
    {
        WRDataR( * pt++);                  //写右半屏数据
    }
}


//设置左半屏起始显示行列地址,x-X 起始行序数(0-7),y-Y 起始列序数(0-63)
void SETXYL(unsigned char x, unsigned char y)
{
    WRComL(x+X);                           //行地址=行序数+行基址
    WRComL(y+Y);                           //列地址=列序数+列基址
}


//设置右半屏起始显示行列地址,x:X 起始行序数(0-7),y:Y 起始列序数(0-63)
void SETXYR(unsigned char x, unsigned char y)
{
    WRComR(x+X);                           //行地址=行序数+行基址
    WRComR(y+Y);                           //列地址=列序数+列基址
}


//****************************************
//显示图形
//****************************************
//显示左半屏一行图形,A-X 起始行序数(0-7),B-Y 起始列地址序数(0-63)
void LineDisL(unsigned char x, unsigned char y, unsigned char * pt)
{
    SETXYL(x,y);                           //设置起始显示行列
    DisplayL(pt, 64);                      //显示数据
```

}

//显示右半屏一行图形,A—X起始行地址序数(0—7),B—Y起始列地址序数(0—63)
void LineDisR(unsigned char x, unsigned char y, unsigned char * pt)
{
    SETXYR(x,y);   //设置起始显示行列
    DisplayR(pt, 64);   //显示数据
}

//***
//显示字体,显示一个数据要占用 X 行两行位置
//***
//右半屏显示一个字节/字:x—起始显示行序数 X(0—7);y—起始显示列序数 Y(0—63);
//pt—显示字数据首地址
void ByteDisR(unsigned char x, unsigned char y,unsigned char * pt)
{
    SETXYR(x,y);   //设置起始显示行列地址
    DisplayR(pt, 8);   //显示上半行数据
    SETXYR(x+1,y);   //设置起始显示行列地址
    DisplayR(pt+8, 8);   //显示下半行数据
}

void WordDisR(unsigned char x, unsigned char y,unsigned char * pt)
{
    SETXYR(x,y);   //设置起始显示行列地址
    DisplayR(pt, 16);   //显示上半行数据
    SETXYR(x+1,y);   //设置起始显示行列地址
    DisplayR(pt+16, 16);   //显示下半行数据
}

//左半屏显示一个字节/字:x—起始显示行序数 X(0—7);y—起始显示列序数 Y(0—63);
//pt—显示字数据首地址
void ByteDisL(unsigned char x, unsigned char y,unsigned char * pt)
{
    SETXYL(x,y);   //设置起始显示行列地址
    DisplayL(pt, 8);   //显示上半行数据
    SETXYL(x+1,y);   //设置起始显示行列地址
    DisplayL(pt+8, 8);   //显示下半行数据
}

void WordDisL(unsigned char x, unsigned char y,unsigned char * pt)
{
    SETXYL(x,y);   //设置起始显示行列地址

```c
        DisplayL(pt, 16);                       //显示上半行数据
        SETXYL(x+1,y);                          //设置起始显示行列地址
        DisplayL(pt+16, 16);                    //显示下半行数据
}
//清屏
void LCDClear()
{
//清左半屏
    unsigned char x,y;
    char j;
    x=0;                                        //起始行,第 0 行
    y=0;                                        //起始列,第 0 列
    for (x=0; x < 8; x++)                       //共 8 行
    {
        SETXYL(x,y);                            //设置起始显示行列地址
        j=64;
        while (j--)
            WRDataL(0);
    }
//清右半屏
    x=0;                                        //起始行,第 0 行
    y=0;                                        //起始列,第 0 列
    for (x=0; x < 8; x++)                       //共 8 行
    {
        SETXYR(x,y);                            //设置起始显示行列地址
        j=64;
        while (j--)
            WRDataR(0);
    }
}

//液晶初始化
void LCD_INIT()
{
    WRComL(0x3e);                               //初始化左半屏,关显示
    WRComL(FirstLine);                          //设置起始显示行,第 0 行
    WRComR(0x3e);                               //初始化右半屏,关显示
    WRComR(FirstLine);                          //设置起始显示行,第 0 行
    LCDClear();                                 //清屏
    WRComL(0x3f);                               //开显示
    WRComR(0x3f);                               //开显示
}
//;主程序 main.c
```

```
//;**************************************************************
//;图形点阵液晶显示器实验说明
//;实验目的:1.掌握使用图形点阵液晶显示字体和图形
//;实验内容:1.显示一个图形
//;         2.显示一段字,包括汉字和英文
//;连线说明:
//;       液晶 12864J:A1 区－－＞A3 区
//;       CS－－＞CS1(0F000H),片选
//;       RW－－＞A0,读/写控制端
//;       RS－－＞A1,数据/指令控制端
//;       CS1/2－－＞A2,左右半屏使能端
//;**************************************************************
extern void LCD_INIT();
extern void WordDisL(unsigned char x, unsigned char y,unsigned char * pt);
extern void WordDisR(unsigned char x, unsigned char y,unsigned char * pt);
extern void ByteDisL(unsigned char x, unsigned char y,unsigned char * pt);
extern void ByteDisR(unsigned char x, unsigned char y,unsigned char * pt);

//--文字: 星  --
code const unsigned char Line1_1[]={
    0x00,0x00,0xFC,0x82,0x82,0xAA,0x2A,0xAA,0xAA,0xAA,0x2A,0x02,0x02,0xFC,0x00,0x00,
    0x00,0xEE,0x9B,0x90,0x98,0x94,0x95,0x80,0x80,0x80,0x95,0x95,0x95,0x95,0xFF,0x00};
//--文字: 研  --
code const unsigned char Line1_2[]={
    0x9E,0x62,0x02,0x02,0x02,0x32,0xFE,0x62,0x02,0x02,0x32,0x02,0x02,0x02,0x62,0xDC,
    0x03,0x3C,0x40,0x40,0x46,0x40,0xF1,0x8E,0x80,0x40,0x7C,0x80,0x80,0x80,0xFE,0x03};
//--文字: 电  --
code const unsigned char Line1_3[]={
    0x00,0xF8,0x04,0x04,0x44,0x44,0x06,0x02,0x02,0x46,0x44,0x04,0x04,0xF8,0x00,0x00,
    0x00,0x0F,0x10,0x10,0x11,0x11,0xF0,0x80,0x90,0x91,0x91,0x8C,0x84,0x87,0xC8,0x78};
//--文字: 子  --
code const unsigned char Line1_4[]={
    0x80,0x40,0x5E,0x52,0x52,0x52,0x32,0x72,0x82,0x82,0x42,0x62,0x52,0x4C,0xC0,0x00,
    0x07,0x04,0x04,0x04,0xFC,0x8C,0x8C,0x80,0x80,0x7C,0x04,0x04,0x04,0x04,0x07,0x00};
//第 2 行显示"星研电子"
void DisLine1()
{
    WordDisL(2,32,Line1_1);          //第 2 行,第 32 列,左半屏,显示一个字子程序
    WordDisL(2,48,Line1_2);
    WordDisR(2,0,Line1_3);           //右半屏,显示一个字子程序
    WordDisR(2,16,Line1_4);
}
```

//"STAR ES51PRO"
```c
code const unsigned char Line2_1[]={
    0x00,0x70,0x88,0x08,0x08,0x08,0x38,0x00,0x00,0x38,0x20,0x21,0x21,0x22,0x1C,0x00};
code const unsigned char Line2_2[]={
    0x18,0x08,0x08,0xF8,0x08,0x08,0x18,0x00,0x00,0x00,0x20,0x3F,0x20,0x00,0x00,0x00};
code const unsigned char Line2_3[]={
    0x00,0x00,0xC0,0x38,0xE0,0x00,0x00,0x00,0x20,0x3C,0x23,0x02,0x02,0x27,0x38,0x20};
code const unsigned char Line2_4[]={
    0x08,0xF8,0x88,0x88,0x88,0x88,0x70,0x00,0x20,0x3F,0x20,0x00,0x03,0x0C,0x30,0x20};
code const unsigned char Line2_5[]={
    0x00,0x00,0x00,0x00,0x00,0x00,0x00,0x00,0x00,0x00,0x00,0x00,0x00,0x00,0x00,0x00};
code const unsigned char Line2_6[]={
    0x08,0xF8,0x88,0x88,0xE8,0x08,0x10,0x00,0x20,0x3F,0x20,0x20,0x23,0x20,0x18,0x00};
code const unsigned char Line2_7[]={
    0x00,0x70,0x88,0x08,0x08,0x08,0x38,0x00,0x00,0x38,0x20,0x21,0x21,0x22,0x1C,0x00};
code const unsigned char Line2_8[]={
    0x00,0xF8,0x08,0x88,0x88,0x08,0x08,0x00,0x00,0x19,0x21,0x20,0x20,0x11,0x0E,0x00};
code const unsigned char Line2_9[]={
    0x00,0x10,0x10,0xF8,0x00,0x00,0x00,0x00,0x00,0x20,0x20,0x3F,0x20,0x20,0x00,0x00};
code const unsigned char Line2_10[]={
    0x08,0xF8,0x08,0x08,0x08,0x08,0xF0,0x00,0x20,0x3F,0x21,0x01,0x01,0x01,0x00,0x00};
code const unsigned char Line2_11[]={
    0x08,0xF8,0x88,0x88,0x88,0x88,0x70,0x00,0x20,0x3F,0x20,0x00,0x03,0x0C,0x30,0x20};
code const unsigned char Line2_12[]={
    0xE0,0x10,0x08,0x08,0x08,0x10,0xE0,0x00,0x0F,0x10,0x20,0x20,0x20,0x10,0x0F,0x00};
//第3行显示"STAR ES51PRO"
void DisLine2()
{
    ByteDisL(4,16,Line2_1);       //第4行,第16列,左半屏,显示一个字节子程序
    ByteDisL(4,24,Line2_2);
    ByteDisL(4,32,Line2_3);
    ByteDisL(4,40,Line2_4);
    ByteDisL(4,48,Line2_5);
    ByteDisL(4,56,Line2_6);

    ByteDisR(4,0,Line2_7);        //右半屏字节显示数据
    ByteDisR(4,8,Line2_8);
    ByteDisR(4,16,Line2_9);
    ByteDisR(4,24,Line2_10);
    ByteDisR(4,32,Line2_11);
    ByteDisR(4,40,Line2_12);
}
```

//－－文字：欢 －－
const unsigned char Line3_1[]={
    0x14,0x24,0x44,0x84,0x64,0x1C,0x20,0x18,0x0F,0xE8,0x08,0x08,0x28,0x18,0x08,0x00,
    0x20,0x10,0x4C,0x43,0x43,0x2C,0x20,0x10,0x0C,0x03,0x06,0x18,0x30,0x60,0x20,0x00};
//－－文字：迎 －－
const unsigned char Line3_2[]={
    0x40,0x41,0xCE,0x04,0x00,0xFC,0x04,0x02,0x02,0xFC,0x04,0x04,0x04,0xFC,0x00,0x00,
    0x40,0x20,0x1F,0x20,0x40,0x47,0x42,0x41,0x40,0x5F,0x40,0x42,0x44,0x43,0x40,0x00};
//－－文字：使 －－
const unsigned char Line3_3[]={
    0x40,0x20,0xF0,0x1C,0x07,0xF2,0x94,0x94,0x94,0xFF,0x94,0x94,0x94,0xF4,0x04,0x00,
    0x00,0x00,0x7F,0x00,0x40,0x41,0x22,0x14,0x0C,0x13,0x10,0x30,0x20,0x61,0x20,0x00};
//－－文字：用 －－
const unsigned char Line3_4[]={
    0x00,0x00,0x00,0xFE,0x22,0x22,0x22,0x22,0xFE,0x22,0x22,0x22,0x22,0xFE,0x00,0x00,
    0x80,0x40,0x30,0x0F,0x02,0x02,0x02,0x02,0xFF,0x02,0x02,0x42,0x82,0x7F,0x00,0x00};

//第 4 行显示"欢迎使用"
void DisLine3()
{
    WordDisL(6,32,Line3_1);   //第 6 行，第 32 列，左半屏，显示一个字子程序
    WordDisL(6,48,Line3_2);   //第 6 行，第 48 列
    WordDisR(6,0,Line3_3);   //右半屏，显示一个字子程序
    WordDisR(6,16,Line3_4);
}

//延时程序
void DelayTime()
{
    unsigned char i;
    unsigned int j;
    for(i=0; i<3; i++)
    {
        for(j=0; j<0xffff; j++)
        {;}
    }
}
```

## 七、实验扩展及思考

修改程序显示"合肥师范欢迎您"并显示设计者的学号和姓名。
```

输出接口仿真实验

一、实验目的

采用 1602 型 LCD 循环显示字符串"Welcome to Heifei Normal University"。其中 LCD 显示模式为：16×2 显示、5×7 点阵、8 位数据口；显示开、有光标并且光标闪烁；光标右移，字符不移。

二、实验原理图

实验参考电路图如图 3-14 所示。

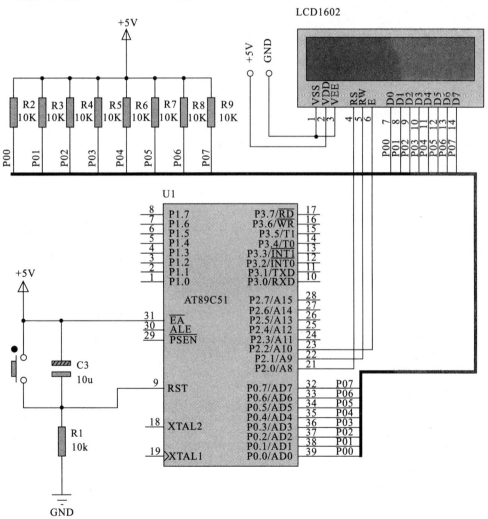

图 3-14 LCD1602 输出接口

三、参考实验程序

```c
//用LCD循环右移显示"Welcome to Heifei Normal University"
#include<reg51.h>            //包含单片机寄存器的头文件
#include<intrins.h>          //包含_nop_()函数定义的头文件
sbit RS=P2^0;                //寄存器选择位,将RS位定义为P2.0引脚
sbit RW=P2^1;                //读写选择位,将RW位定义为P2.1引脚
sbit E=P2^2;                 //使能信号位,将E位定义为P2.2引脚
sbit BF=P0^7;                //忙碌标志位,将BF位定义为P0.7引脚
unsigned char code string[ ]={"Welcome toHeifei Normal University"};

void delay1ms()              //延时函数1ms
{
    unsigned char i,j;
    for(i=0;i<10;i++)
        for(j=0;j<33;j++)
            ;
}

void delay(unsigned char n)  //延时函数nms
{
    unsigned char i;
    for(i=0;i<n;i++)
        delay1ms();
}

unsigned char BusyTest(void) //忙闲检测:返回值result=1,忙;result=0,闲
{
    bit result;
    RS=0;                    //根据规定,RS为低电平,RW为高电平时,可以读状态
    RW=1;
    E=1;                     //E=1,才允许读写
    _nop_();                 //空操作
    _nop_();
    _nop_();
    _nop_();                 //空操作4个机器周期,给硬件反应时间
    result=BF;               //将忙碌标志电平赋给result
    E=0;
    return result;
}
/************写命令函数*************/
void WriteInstruction (unsigned char dictate)
{
    while(BusyTest()==1);
```

```c
    RS=0;
    RW=0;
    E=0;                    //E置低电平(写指令时,E为高脉冲,
                            //就是让E从0到1发生正跳变,所以应先置"0"
    _nop_();
    _nop_();                //空操作2个机器周期,给硬件反应时间
    P0=dictate;             //将数据送入P0口,即写入指令或地址
    _nop_();
    _nop_();
    _nop_();
    _nop_();                //空操作4个机器周期,给硬件反应时间
    E=1;                    //E置高电平
    _nop_();
    _nop_();
    _nop_();
    _nop_();                //空操作4个机器周期,给硬件反应时间
    E=0;                    //当E由高电平跳变成低电平时,液晶模块开始执行命令
}
/*********写地址函数:地址为x,**********/
void WriteAddress(unsigned char x)
{
    WriteInstruction(x|0x80);    //显示位置的确定方法规定为"80H+地址码x"
}
/********写数据函数,数据为字符y**************/
void WriteData(unsigned char y)
{
    while(BusyTest()==1);
    RS=1;                   //RS为高电平,RW为低电平时,可以写入数据
    RW=0;
    E=0;                    //E置低电平(写指令时,E为高脉冲,
                            //就是让E从0到1发生正跳变,所以应先置"0"
    P0=y;                   //将数据送入P0口,即将数据写入液晶模块
    _nop_();
    _nop_();
    _nop_();
    _nop_();                //空操作4个机器周期,给硬件反应时间
    E=1;                    //E置高电平
    _nop_();
    _nop_();
    _nop_();
    _nop_();                //空操作4个机器周期,给硬件反应时间
    E=0;                    //当E由高电平跳变成低电平时,液晶模块开始执行命令
}
```

```c
    void LcdInitiate(void)           //Lcd 初始化函数
    {
        delay(15);                   //延时 15ms
        WriteInstruction(0x38);      //16×2 显示,5×7 点阵,8 位数据接口
        delay(5);                    //延时 5ms
        WriteInstruction(0x38);
        delay(5);
        WriteInstruction(0x38);
        delay(5);
        WriteInstruction(0x0f);      //显示模式设置:显示开,有光标,光标闪烁
        delay(5);
        WriteInstruction(0x06);      //显示模式设置:光标右移,字符不移
        delay(5);
        WriteInstruction(0x01);      //清屏幕指令,将以前的显示内容清除
        delay(5);
    }
    void main(void)                  //主函数
    {
        unsigned char i;
        LcdInitiate();               //调用 LCD 初始化函数
        delay(10);
        while(1)
        {
            WriteInstruction(0x01);  //清显示:清屏幕指令
            WriteAddress(0x00);      //设置显示位置为第一行的第 5 个字
            i=0;
            while(string[i]!='\0')
            {   WriteData(string[i]);
                i++;
                delay(150);
            }
            for(i=0;i<4;i++)
                delay(250);
        }
    }
```

四、实验思考题

试编写 LCD12864 显示初始化程序,设计电路,采用汉字取模,使 LCD 循环显示字符串"hefei normal university"和个人学号、姓名。

要求:字符分为两行,居中显示,字符从左向右缓慢移动。

实验 7　综合设计实验(定时器、中断综合实验——电子钟)

一、实验目的

熟悉 MCS51 类 CPU 的定时器、中断系统编程方法,了解定时器的应用、实时程序的设计和调试技巧。

二、实验内容

编写一个时钟程序,使用定时器产生一个 50ms 的定时中断,对定时中断计数,将时、分、秒显示在数码管上。

三、程序框图

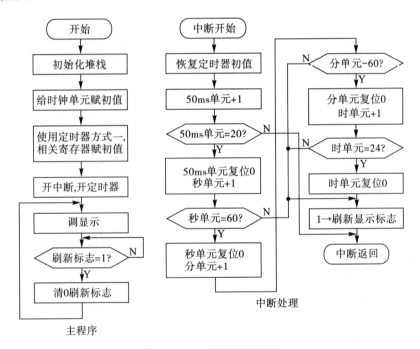

图 3-15　电子钟程序流程图

四、实验步骤

1. 连线说明:

E5 区 :CLK——B2 区 :2M

E5 区 :CS——A3 区 :CS5

E5 区 :A0——A3 区 :A0

E5 区 :A、B、C、D——G5 区 :A、B、C、D

2. 时间显示在数码管上。

五、程序清单

```
ms50        DATA        31H             ;存放多少个50ms
sec         DATA        32H             ;秒
min         DATA        33H             ;分
hour        DATA        34H             ;时
buffer      DATA        35H             ;显示缓冲区
EXTRNCODE(Display8)
            ORG         0000H
            LJMP        STAR
            ORG         000BH           ;定时器T0中断处理入口地址
            LJMP        INT_Timer0
            ORG         0100H
STAR:       MOV         SP,#60H         ;堆栈
            MOV         ms50,A          ;清零ms50
            MOV         hour,#12        ;设定初值:12:59:50
            MOV         min,#59
            MOV         sec,#50
            MOV         TH0,#60         ;定时中断计数器初值
            MOV         TL0,#176        ;定时50ms
            MOV         TMOD,#1         ;定时器0:方式一
            MOV         IE,#82H         ;允许定时器0中断
            SETB        TR0             ;开定时器T0
STAR1:      LCALL       Display         ;调用显示
            JNB         F0,$
            CLR         F0
            SJMP        STAR1           ;需要重新显示时间;中断服务程序
INT_Timer0: MOV         TL0,#176-5
            MOV         TH0,#60
            PUSH        01H
            MOV         R1,#ms50
            INC         @R1             ;50ms单元加1
            CJNE        @R1,#20,ExitInt
            MOV         @R1,#0          ;恢复初值
            INC         R1
            INC         @R1             ;秒加1
            CJNE        @R1,#60,ExitInt1
            MOV         @R1,#0
            INCR1
            INC         @R1             ;分加1
            CJNE        @R1,#60,ExitInt1
            MOV         @R1,#0
            INCR1
```

```
            INC         @R1              ;时加1
            CJNE        @R1,#24,ExitInt1
            MOV         @R1,#0
ExitInt1:   SETB        F0
ExitInt:    POP         01H
            RETI
HexToBCD:   MOV         B,#10
            DIV         AB
            MOV         @R0,B
            INC         R0
            MOV         @R0,A
            INCR0
            RET
Display:    MOV         R0,#buffer
            MOV         A,sec
            ACALL       HexToBCD
            MOV         @R0,#10H         ;第三位不显示
            INCR0
            MOV         A,min
            ACALLHexToBCD
            MOV         @R0,#10H         ;第六位不显示
            INC         R0
            MOV         A,hour
            ACALL       HexToBCD
            MOV         R0,#buffer
            LCALL       Display8
            RET
            END
```

六、思考题

1. 电子钟走时精度与哪些有关系？中断程序中给 TL0 赋值为什么与初始化程序中不一样？
2. 使用 C51 语言编写电子时钟程序。

仿真实验综合设计——案例1

基于 AT89C52 单片机的篮球赛计分器设计

一、实验目的

采用单片机 AT89C52 作为控制芯片,利用 LCD12864 作为显示器件,设计一个篮球比赛计分器系统,实现能记录、修改、暂停整个赛程的比赛时间,能随时刷新甲、乙两队在整个赛程中的比分,中场交换比赛场地时能两队比分,比赛结束能发出报警指令。

二、实验原理图

1. 应用系统设计

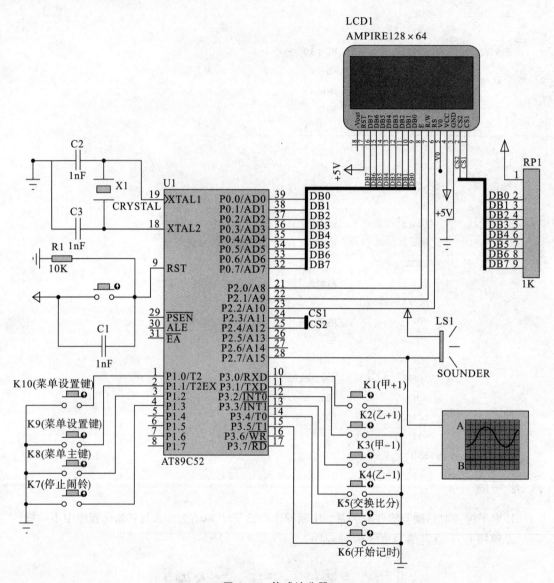

图 3-16　篮球计分器

2. 接口设计

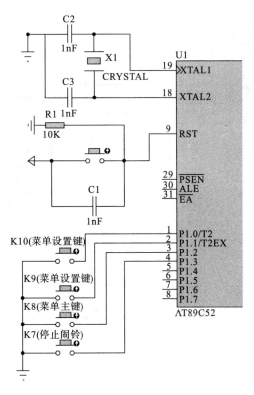

图 3-17　篮球计分器接口

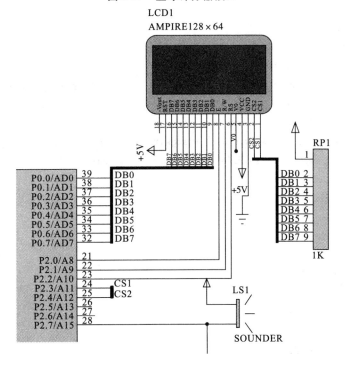

图 3-18　LCD12864 接口图

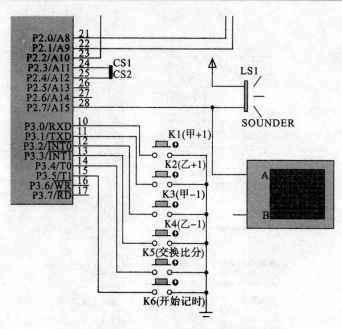

图 3-19 蜂鸣器及键盘接口

三、参考实验程序

```c
#include <reg52.h>
#include <intrins.h>
#include "12864drive.h"
#include "calendar.h"
#define uint unsigned int
#define uchar unsigned char
uchar aa,y,niangw=20,niandw=11,yue=5,ri=10,shi=00,fen=00,miao=00,bifen1=00,bifen2=00;
uchar nianqw,nianbw,niansw,niangew,yuesw,yuegw,risw,rigw;
uchar shisw,shigw,fensw,fengw,miaosw,miaogw,ringshi=00,ringfen=01,ringshisw,ringshigw,ringfensw,ringfengw,bifen1sw,bifen1gw,bifen2sw,bifen2gw;
uchar keynum,yuezuidashu;
void Displaystart();
sbit        KUP=P1^0;
sbit        KDOWN=P1^1;
sbit        KEY=P1^2;
sbit        BELL=P2^7;
sbit        F1=P1^3;
sbit        A1=P3^0;
sbit        B1=P3^1;
sbit        C1=P3^2;
sbit        D1=P3^3;
sbit        E1=P3^4;
```

```c
sbit        G1=P3^5;
bit         b_close0=1;
bit         b_close1=1;
bit         keyf;           //设置键是否按下标志位
bit         kupf;           //加一键是否按下标志位
bit         kdownf;         //减一键是否按下标志位
bit         ykeyf;          //设置键消抖动标志位
bit         ykupf;          //加一键消抖动标志位
bit         ykdownf;        //减一键消抖动标志位
bit         a1;
bit         b1;
bit         c1;
bit         d1;
bit         e1;
bit         f1;
bit         g1;
bit         show;
bit         b_bell=0;
bit         b_ring=1;       //闹铃开启标志位

void jxchengxu()
{
    nianqw=niangw/10;
    nianbw=niangw%10;
    niansw=niandw/10;
    niangew=niandw%10;
    yuesw=yue/10;
    yuegw=yue%10;
    risw=ri/10;
    rigw=ri%10;
    shisw=shi/10;
    shigw=shi%10;
    fensw=fen/10;
    fengw=fen%10;
    miaosw=miao/10;
    miaogw=miao%10;
    bifen1sw=bifen1/10;
    bifen1gw=bifen1%10;
    bifen2sw=bifen2/10;
    bifen2gw=bifen2%10;
}

void delay()
```

```c
{
    uchar i,j,k;
    for(i=0;i<255;i++)
        for(j=0;j<255;j++)
            for(k=0;k<5;k++);
}

void SystemInit()
{
    TMOD=0x01;
    TH0=(65535-40000)/256;
    TL0=(65535-40000)%256;
    EA=1;
    ET0=1;
    TR0=1;
}
void jianpan()              //键盘扫描子程序
{
    P3=P3|0xb8;
    if(KEY==0)
    {
        keyf=1;
        return;
    }
    if(KDOWN==0)
    {
        kdownf=1;
        return;
    }
    if(KUP==0)
    {
        kupf=1;
        return;
    }
    if(A1==0)
    {
        a1=1;
        return;
    }
    if(B1==0)
    {
        b1=1;
        return;
```

```
    }
    if(C1==0)
    {
        c1=1;
        return;
    }
    if(D1==0)
    {
        d1=1;
        return;
    }
    if(E1==0)
    {
        e1=1;
        return;
    }
    if(F1==0)
    {
        f1=1;
        return;
    }
    if(G1==0)
    {
        g1=1;
        return;
    }
}

void DisplayMenu0()
{
    ClearScreen(0);                          //清屏
    Display(2,0,0,16,1);                     //时间
    Display(2,0,16,17,1);
    Display(2,0,32,14,1);                    //显示"设置"两字
    Display(2,0,48,15,1);
    Display(2,2,0,18,0);                     //时期
    Display(2,2,16,19,0);
    Display(2,2,32,14,0);                    //显示"设置"两字
    Display(2,2,48,15,0);
    Display(2,4,0,20,0);
    Display(2,4,16,21,0);
    Display(2,4,32,14,0);
    Display(2,4,48,15,0);
```

```
        Display(2,6,0,24,0);
        Display(2,6,16,25,0);
}

void DisplayMenu1()
{
        ClearScreen(0);                         //清屏
        Display(2,0,0,16,0);
        Display(2,0,16,17,0);
        Display(2,0,32,14,0);                   //显示"设置"两字
        Display(2,0,48,15,0);
        Display(2,2,0,18,1);
        Display(2,2,16,19,1);
        Display(2,2,32,14,1);                   //显示"设置"两字
        Display(2,2,48,15,1);
        Display(2,4,0,20,0);
        Display(2,4,16,21,0);
        Display(2,4,32,14,0);
        Display(2,4,48,15,0);
        Display(2,6,0,24,0);
        Display(2,6,16,25,0);
}

void DisplayMenu2()
{
        ClearScreen(0);                         //清屏
        Display(2,0,0,16,0);
        Display(2,0,16,17,0);
        Display(2,0,32,14,0);                   //显示"设置"两字
        Display(2,0,48,15,0);
        Display(2,2,0,18,0);
        Display(2,2,16,19,0);
        Display(2,2,32,14,0);                   //显示"设置"两字
        Display(2,2,48,15,0);
        Display(2,4,0,20,1);
        Display(2,4,16,21,1);
        Display(2,4,32,14,1);
        Display(2,4,48,15,1);
        Display(2,6,0,24,0);
        Display(2,6,16,25,0);
}
void keychengxu()
{
```

```
if(ykeyf==1)
    return;
ykeyf=1;
if((b_bell==1)&&(b_close0==1)&&b_ring==1)    //如此时正响铃,则关闭铃声并退出
{
    b_close0=0;
    return;
}
if(keynum==0)
{
    show=0;
    keynum=1;
    Displayen(2,6,8,14,0);
    Displayen(2,6,40,15,0);
    Display(2,6,8,14,1);
    Display(2,6,24,15,1);
    return;
}
if(keynum==1)
{
    show=1;
    keynum=2;
    DisplayMenu0();
    return;
}
if(keynum==2)
{
    show=1;
    keynum=3;
    TR0=0;
    ClearScreen(0);                                  //清屏
    Display(2,0,32,16,0);
    Display(2,0,48,17,0);
    Display(1,0,0,14,0);
    Display(1,0,16,15,0);
    Displayen(2,4,48,10,0);
    Displayen(1,4,8,10,0);
    Displayen(2,4,32,shisw,0);
    Displayen(2,4,40,shigw,0);
    Displayen(2,4,56,fensw,0);
    Displayen(1,4,0,fengw,0);
    Displayen(1,4,16,miaosw,1);
    Displayen(1,4,24,miaogw,1);
```

```c
        return;
    }
    if(keynum==3)
    {
        keynum=5;
        Displayen(2,4,56,fensw,1);
        Displayen(1,4,0,fengw,1);
        Displayen(1,4,16,miaosw,0);
        Displayen(1,4,24,miaogw,0);
        return;
    }
    if(keynum==4)
    {
        show=1;
        ClearScreen(0);                         //清屏
        Display(2,0,32,18,0);
        Display(2,0,48,19,0);
        Display(1,0,0,14,0);
        Display(1,0,16,15,0);
        Display(2,4,2*16,7,0);                  //显示年
        Display(1,4,0*16,8,0);                  //显示月
        Display(1,4,2*16,9,0);                  //显示日
        Displayen(2,4,0,nianqw,0);
        Displayen(2,4,8,nianbw,0);
        Displayen(2,4,16,niansw,0);
        Displayen(2,4,24,niangew,0);
        Displayen(2,4,48,yuesw,0);
        Displayen(2,4,56,yuegw,0);
        Displayen(1,4,16,risw,1);
        Displayen(1,4,24,rigw,1);
        keynum=7;
        return;
    }
    if(keynum==7)
    {
        Displayen(2,4,48,yuesw,1);
        Displayen(2,4,56,yuegw,1);
        Displayen(1,4,16,risw,0);
        Displayen(1,4,24,rigw,0);
        keynum=8;
        return;
    }
    if(keynum==8)
```

```c
{
    Displayen(2,4,0,nianqw,1);
    Displayen(2,4,8,nianbw,1);
    Displayen(2,4,16,niansw,1);
    Displayen(2,4,24,niangew,1);
    Displayen(2,4,48,yuesw,0);
    Displayen(2,4,56,yuegw,0);
    keynum=9;
    return;
}
if(keynum==9)
{
    keynum=4;
    DisplayMenu1();
    return;
}
if(keynum==5)
{
    keynum=6;
    Displayen(2,4,56,fensw,0);
    Displayen(1,4,0,fengw,0);
    Displayen(2,4,32,shisw,1);
    Displayen(2,4,40,shigw,1);
    return;
}
if(keynum==6)
{
    TR0=1;                              //时间设置马上计时
    keynum=2;
    DisplayMenu0();
    return;
}
if(keynum==10)
{
    ClearScreen(0);                     //清屏
    keynum=14;
    Display(1,4,48,27,0);               //显示闹铃图标
    Displayen(1,4,16,14,0);
    if(b_ring==1)
    Display(1,4,24,22,0);               //显示"开"字
    else
    Display(1,4,24,23,0);               //显示"关"字
    Display(1,6,24,26,0);
```

```
        Displayen(1,4,40,15,0);
        Display(2,0,32,20,0);
        Display(2,0,48,21,0);
        Display(1,0,0,14,0);
        Display(1,0,16,15,0);
        Displayen(2,4,32,ringshisw,0);
        Displayen(2,4,40,ringshigw,0);
        Displayen(2,4,56,ringfensw,1);
        Displayen(1,4,0, ringfengw,1);
        Displayen(2,4,48,10,0);              //显示冒号
        return;
    }
    if(keynum==14)
    {
        Displayen(2,4,32,ringshisw,1);
        Displayen(2,4,40,ringshigw,1);
        Displayen(2,4,56,ringfensw,0);
        Displayen(1,4,0, ringfengw,0);
        keynum=15;
        return;
    }
    if(keynum==15)
    {
        if(b_ring==1)
            Display(1,4,24,22,1);            //显示"开"字
        else
            Display(1,4,24,23,1);            //显示"关"字
        keynum=16;
        return;
    }
    if(keynum==16)
    {
        keynum=10;
        DisplayMenu2();
        return;
    }
    if(keynum==11)
    {
        show=0;
        keynum=0;
        ClearScreen(0);                      //清屏
        Displaystart();
    }
```

```c
}
void kdownchengxu()
{
    if(ykdownf==1)
        return;
    ykdownf=1;
    if((b_bell==1)&&(b_close0==1)&&b_ring==1)    //如此时正响铃则关闭铃声并退出
    {
        b_close0=0;
        return;
    }
    if(keynum==2)
    {
        Display(2,0,0,16,0);                      //显示"时间"两字
        Display(2,0,16,17,0);
        Display(2,0,32,14,0);                     //显示"设置"两字
        Display(2,0,48,15,0);
        Display(2,2,0,18,1);                      //显示"日期"两字
        Display(2,2,16,19,1);
        Display(2,2,32,14,1);                     //显示"设置"两字
        Display(2,2,48,15,1);
        Display(2,4,0,20,0);
        Display(2,4,16,21,0);
        Display(2,4,32,14,0);
        Display(2,4,48,15,0);
        Display(2,6,0,24,0);                      //显示"返回"两字
        Display(2,6,16,25,0);
        keynum=4;
        return;
    }
    if(keynum==4)                                 //日期设置状态
    {
        Display(2,0,0,16,0);
        Display(2,0,16,17,0);
        Display(2,0,32,14,0);                     //显示"设置"两字
        Display(2,0,48,15,0);
        Display(2,2,0,18,0);
        Display(2,2,16,19,0);
        Display(2,2,32,14,0);                     //显示"设置"两字
        Display(2,2,48,15,0);
        Display(2,4,0,20,1);
        Display(2,4,16,21,1);
        Display(2,4,32,14,1);
```

```
        Display(2,4,48,15,1);
        Display(2,6,0,24,0);
        Display(2,6,16,25,0);
        keynum=10;
        return;
}
if(keynum==10)                            //日期设置状态
{
        Display(2,0,0,16,0);
        Display(2,0,16,17,0);
        Display(2,0,32,14,0);              //显示"设置"两字
        Display(2,0,48,15,0);
        Display(2,2,0,18,0);
        Display(2,2,16,19,0);
        Display(2,2,32,14,0);              //显示"设置"两字
        Display(2,2,48,15,0);
        Display(2,4,0,20,0);
        Display(2,4,16,21,0);
        Display(2,4,32,14,0);
        Display(2,4,48,15,0);
        Display(2,6,0,24,1);
        Display(2,6,16,25,1);
        keynum=11;
        return;
}
if(keynum==11)
{
        Display(2,0,0,16,1);
        Display(2,0,16,17,1);
        Display(2,0,32,14,1);              //显示"设置"两字
        Display(2,0,48,15,1);
        Display(2,2,0,18,0);
        Display(2,2,16,19,0);
        Display(2,2,32,14,0);              //显示"设置"两字
        Display(2,2,48,15,0);
        Display(2,4,0,20,0);
        Display(2,4,16,21,0);
        Display(2,4,32,14,0);
        Display(2,4,48,15,0);
        Display(2,6,0,24,0);
        Display(2,6,16,25,0);
        keynum=2;
}
```

```
if(keynum==3)
{
    miao--;
    if(miao==-1)
        miao=59;
    miaosw=miao/10;
    miaogw=miao%10;
    Displayen(1,4,16,miaosw,1);
    Displayen(1,4,24,miaogw,1);
}
if(keynum==5)
{
    fen--;
    if(fen==-1)
        fen=59;
    fensw=fen/10;
    fengw=fen%10;
    Displayen(2,4,56,fensw,1);
    Displayen(1,4,0,fengw,1);
}
if(keynum==6)
{
    shi--;
    if(shi==-1)
        shi=23;
    shisw=shi/10;
    shigw=shi%10;
    Displayen(2,4,32,shisw,1);
    Displayen(2,4,40,shigw,1);
}
if(keynum==7)
{
    ri--;
    if(yue==2&&ri==0)
        ri=month_n_day((niangw<<6)+(niangw<<5)+(niangw<<2)+niandw,2);
    if((yue==1||yue==3||yue==5||yue==7||yue==8||yue==10||yue==12)&&ri==0)
        ri=31;
    if((yue==4||yue==6||yue==9||yue==11)&&ri==00)
        ri=30;
    risw=ri/10;
    rigw=ri%10;
    Displayen(1,4,16,risw,1);
    Displayen(1,4,24,rigw,1);
```

```c
    }
    if(keynum==8)
    {
        yue--;
        if(yue==0)
            yue=12;
        yuesw=yue/10;
        yuegw=yue%10;
yuezuidashu=month_n_day((niangw<<6)+(niangw<<5)+(niangw<<2)+niandw,yue);
        if(ri>yuezuidashu)
        {
            ri=yuezuidashu;
            risw=ri/10;
            rigw=ri%10;
            Displayen(1,4,16,risw,0);
            Displayen(1,4,24,rigw,0);
        }
        Displayen(2,4,48,yuesw,1);
        Displayen(2,4,56,yuegw,1);
    }
    if(keynum==9)
    {
        niandw--;
        if(niandw==-1)
        {
            niandw=99;
            niangw--;
        }
        nianqw=niangw/10;
        nianbw=niangw%10;
        niansw=niandw/10;
        niangew=niandw%10;
        Displayen(2,4,0,nianqw,1);
        Displayen(2,4,8,nianbw,1);
        Displayen(2,4,16,niansw,1);
        Displayen(2,4,24,niangew,1);
    }
    if(keynum==14)
    {
        ringfen--;
        if(ringfen==-1)
            ringfen=59;
        ringfensw=ringfen/10;
```

```c
        ringfengw=ringfen%10;
        Displayen(2,4,56,ringfensw,1);
        Displayen(1,4,0, ringfengw,1);
    }
    if(keynum==15)
    {
        ringshi--;
        if(ringshi==-1)
            ringshi=23;
        ringshisw=ringshi/10;
        ringshigw=ringshi%10;
        Displayen(2,4,32,ringshisw,1);
        Displayen(2,4,40,ringshigw,1);
    }
    if(keynum==16)
    {
        b_ring=! b_ring;
        if(b_ring==1)
            Display(1,4,24,22,1);                 //显示"开"字
        else
            Display(1,4,24,23,1);                 //显示"关"字
        keynum=16;
    }
}

void Displaystart()
{
    Display(2,2,40,7,0);                          //显示年
    Display(1,2,8,8,0);                           //显示月
    Display(1,2,40,9,0);                          //显示日
    Displayen(1,0,0,12,0);
    Displayen(1,0,8,13,0);
    Display(1,0,16,9,0);
    Display(1,0,32,10,0);
    Display(1,0,48,11,0);
    Display(2,0,16,12,0);
    Display(2,0,32,13,0);
    Display(2,0,48,11,0);
    Displayen(2,6,0,14,0);
    Displayen(2,6,40,15,0);
    Display(2,6,8,5,0);                           //显示"比分"两字
    Display(2,6,24,6,0);
    Display(1,6,48,4,0);
```

```
        Displayen(2,4,48,10,0);
        Displayen(1,4,8,10,0);
}

    void jiaohuan()
    {
        delay();
        if(e1==1)
        {
        int temp;
        temp=bifen1;
        bifen1=bifen2;
        bifen2=temp;
        Displayen(1,6,0,bifen1sw,0);
        Displayen(1,6,8,bifen1gw,0);
        Displayen(1,6,24,bifen2sw,0);
        Displayen(1,6,32,bifen2gw,0);
        Displayen(1,0,0,12,0);
        Displayen(1,0,8,13,0);
        Display(2,0,16,9,0);
        Display(2,0,32,10,0);
        Display(2,0,48,11,0);
        Display(1,0,16,12,0);
        Display(1,0,32,13,0);
        Display(1,0,48,11,0);

        }
    }

    void tingzhi()
    {
        if((b_bell==1)&&(b_close0==1)&&b_ring==1)    //如此时正响铃则关闭铃声并退出
        {
            b_close0=0;
            return;
        }
    }

    void jiafen1()
    {
        delay();
        if(a1==1)
        {   bifen1++;
```

```c
        if(bifen1==100)
            bifen1=0;
        bifen1sw=bifen1/10;
        bifen1gw=bifen1 % 10;
        Displayen(1,6,0,bifen1sw,0);
        Displayen(1,6,8,bifen1gw,0);
    }
}
void kaishi()
{
    delay();
    if(g1==1)
    {   EA=1;
    }
}


void jiafen2()
{
    delay();
    if(b1==1)
    {   bifen2++;
        if(bifen2==100)
            bifen2=0;
        bifen2sw=bifen2/10;
        bifen2gw=bifen2 % 10;
        Displayen(1,6,24,bifen1sw,0);
        Displayen(1,6,32,bifen1gw,0);
    }
}

void jianfen1()
{
    delay();
    if(c1==1)
    {   bifen1--;
        if(bifen1==100)
            bifen1=0;
        bifen1sw=bifen1/10;
        bifen1gw=bifen1 % 10;
        Displayen(1,6,0,bifen1sw,0);
        Displayen(1,6,8,bifen1gw,0);
    }
}
```

```
void jianfen2()
{
    delay();
    if(d1==1)
    {   bifen2--;
        if(bifen2==100)
            bifen2=0;
        bifen2sw=bifen2/10;
        bifen2gw=bifen2%10;
        Displayen(1,6,24,bifen2sw,0);
        Displayen(1,6,32,bifen2gw,0);
    }
}
void kupchengxu()
{
    if(ykupf==1)
        return;
    ykupf=1;
    if((b_bell==1)&&(b_close0==1)&&b_ring==1)       //如此时正响铃则关闭铃声并退出
    {
        b_close0=0;
        return;
    }
    if(keynum==2)
    {
        Display(2,0,0,16,0);
        Display(2,0,16,17,0);
        Display(2,0,32,14,0);                        //显示"设置"两字
        Display(2,0,48,15,0);
        Display(2,2,0,18,0);
        Display(2,2,16,19,0);
        Display(2,2,32,14,0);                        //显示"设置"两字
        Display(2,2,48,15,0);
        Display(2,4,0,20,0);
        Display(2,4,16,21,0);
        Display(2,4,32,14,0);
        Display(2,4,48,15,0);
        Display(2,6,0,24,1);
        Display(2,6,16,25,1);
        keynum=11;
        return;
    }
```

```
if(keynum==11)                                      //日期设置状态
{
    Display(2,0,0,16,0);
    Display(2,0,16,17,0);
    Display(2,0,32,14,0);                           //显示"设置"两字
    Display(2,0,48,15,0);
    Display(2,2,0,18,0);
    Display(2,2,16,19,0);
    Display(2,2,32,14,0);                           //显示"设置"两字
    Display(2,2,48,15,0);
    Display(2,4,0,20,1);
    Display(2,4,16,21,1);
    Display(2,4,32,14,1);
    Display(2,4,48,15,1);
    Display(2,6,0,24,0);
    Display(2,6,16,25,0);
    keynum=10;
    return;
}
if(keynum==10)                                      //日期设置状态
{
    Display(2,0,0,16,0);
    Display(2,0,16,17,0);
    Display(2,0,32,14,0);                           //显示"设置"两字
    Display(2,0,48,15,0);
    Display(2,2,0,18,1);
    Display(2,2,16,19,1);
    Display(2,2,32,14,1);                           //显示"设置"两字
    Display(2,2,48,15,1);
    Display(2,4,0,20,0);
    Display(2,4,16,21,0);
    Display(2,4,32,14,0);
    Display(2,4,48,15,0);
    Display(2,6,0,24,0);
    Display(2,6,16,25,0);
    keynum=4;
    return;
}
if(keynum==4)
{
    Display(2,0,0,16,1);
    Display(2,0,16,17,1);
    Display(2,0,32,14,1);                           //显示"设置"两字
```

```
        Display(2,0,48,15,1);
        Display(2,2,0,18,0);
        Display(2,2,16,19,0);
        Display(2,2,32,14,0);                //显示"设置"两字
        Display(2,2,48,15,0);
        Display(2,4,0,20,0);
        Display(2,4,16,21,0);
        Display(2,4,32,14,0);
        Display(2,4,48,15,0);
        Display(2,6,0,24,0);
        Display(2,6,16,25,0);
        keynum=2;
    }
    if(keynum==3)
    {
        miao++;
        if(miao==60)
        miao=0;
        miaosw=miao/10;
        miaogw=miao%10;
        Displayen(1,4,16,miaosw,1);
        Displayen(1,4,24,miaogw,1);
    }
    if(keynum==5)
    {
        fen++;
        if(fen==60)
        fen=0;
        fensw=fen/10;
        fengw=fen%10;
        Displayen(2,4,56,fensw,1);
        Displayen(1,4,0,fengw,1);
    }
    if(keynum==6)
    {
        shi++;
        if(shi==24)
            shi=0;
        shisw=shi/10;
        shigw=shi%10;
        Displayen(2,4,32,shisw,1);
        Displayen(2,4,40,shigw,1);
    }
```

```c
if(keynum==7)
{
    ri++;
    if(yue==2)
    {
    yuezuidashu=month_n_day((niangw<<6)+(niangw<<5)+(niangw<<2)+niandw,2);
        if(ri>yuezuidashu)
            ri=1;
    }
    if((yue==1||yue==3||yue==5||yue==7||yue==8||yue==10||yue==12)&&ri==32)
        ri=1;
    if((yue==4||yue==6||yue==9||yue==11)&&ri==31)
        ri=1;
    risw=ri/10;
    rigw=ri%10;
    Displayen(1,4,16,risw,1);
    Displayen(1,4,24,rigw,1);
}
if(keynum==8)
{
    yue++;
    if(yue==13)
    yue=1;
    yuezuidashu=month_n_day((niangw<<6)+(niangw<<5)+(niangw<<2)+niandw,yue);
    if(ri>yuezuidashu)
    {
        ri=yuezuidashu;
        risw=ri/10;
        rigw=ri%10;
        Displayen(1,4,16,risw,0);
        Displayen(1,4,24,rigw,0);
    }
    yuesw=yue/10;
    yuegw=yue%10;
    Displayen(2,4,48,yuesw,1);
    Displayen(2,4,56,yuegw,1);
}
if(keynum==9)
{
    niandw++;
    if(niandw==100)
    {
        niandw=0;
```

```
            niangw++;
        }
        nianqw=niangw/10;
        nianbw=niangw%10;
        niansw=niandw/10;
        niangew=niandw%10;
        Displayen(2,4,0,nianqw,1);
        Displayen(2,4,8,nianbw,1);
        Displayen(2,4,16,niansw,1);
        Displayen(2,4,24,niangew,1);
    }
    if(keynum==14)
    {
        ringfen++;
        if(ringfen==60)
            ringfen=0;
        ringfensw=ringfen/10;
        ringfengw=ringfen%10;
        Displayen(2,4,56,ringfensw,1);
        Displayen(1,4,0, ringfengw,1);
    }
    if(keynum==15)
    {
        ringshi++;
        if(ringshi==24)
            ringshi=0;
        ringshisw=ringshi/10;
        ringshigw=ringshi%10;
        Displayen(2,4,32,ringshisw,1);
        Displayen(2,4,40,ringshigw,1);
    }
}

void jianzhichuli()                              //键值处理子程序
{
    if(keyf==1)
    {
        if(KEY==0)
            keychengxu();
        else
        {
            ykeyf=0;
            keyf=0;
```

```
        }
    }
    if(kupf==1)
    {
        if(KUP==0)
            kupchengxu();
        else
        {
            ykupf=0;
            kupf=0;
        }
    }
    if(kdownf==1)
    {
        if(KDOWN==0)
            kdownchengxu();
        else
        {
            ykdownf=0;
            kdownf=0;
        }
    }
    if(a1==1)
    {
        if(A1==0)
            jiafen1();
        else
        {
            a1=0;
        }
    }
    if(b1==1)
    {
        if(B1==0)
            jiafen2();
        else
        {
            b1=0;
        }
    }
    if(c1==1)
    {
        if(C1==0)
```

```
                jianfen1();
            else
            {
                c1=0;
            }
        }
        if(d1==1)
        {
            if(D1==0)
                jianfen2();
            else
            {
                c1=0;
            }
        }
        if(e1==1)
        {
            if(E1==0)
                jiaohuan();
            else
            {
                e1=0;
            }
        }
        if(f1==1)
        {
            if(F1==0)
                tingzhi();
            else
            {
                f1=0;
            }
        }
        if(g1==1)
        {
            if(G1==0)
                kaishi();
            else
            {
                g1=0;
            }
        }
    }
```

```
void bellout()
{   uchar i,j;
    for(i=0;i<300;i++)
        for(j=0;j<300;j++)
    BELL=~BELL;
}

void main()
{
    InitLCD();                              //初始12864
    SystemInit();
    ClearScreen(0);                         //清屏
    SetStartLine(0);                        //显示开始行
    Displaystart();
    while(1)
    {
        jxchengxu();
        if(show==0)
        {
            Displayen(2,2,8,nianqw,0);
            Displayen(2,2,16,nianbw,0);
            Displayen(2,2,24,niansw,0);
            Displayen(2,2,32,niangew,0);
            Displayen(2,2,56,yuesw,0);
            Displayen(1,2,0,yuegw,0);
            Displayen(1,2,24,risw,0);
            Displayen(1,2,32,rigw,0);
            Displayen(2,4,32,shisw,0);
            Displayen(2,4,40,shigw,0);
            Displayen(2,4,56,fensw,0);
            Displayen(1,4,0,fengw,0);
            Displayen(1,4,16,miaosw,0);
            Displayen(1,4,24,miaogw,0);
            Displayen(1,6,0,bifen1sw,0);
            Displayen(1,6,8,bifen1gw,0);
            Displayen(1,6,16,10,0);
            Displayen(1,6,24,bifen2sw,0);
            Displayen(1,6,32,bifen2gw,0);
            if((miao>00)&&(fen>=01))
            {   Display(1,6,48,3,0);
            }
            if((fen==01)&&(miao==00))
```

```
            {    Display(1,6,48,2,0);}
                if(b_ring==1)
            {    Display(1,4,48,27,0);}
        }
            jianpan();
            jianzhichuli();
            if((b_bell==1)&&(b_close0==1)&&b_ring==1)
                bellout();                                          //闹铃时间到响铃
            if(b_bell==0)
                b_close0=1;
    }
}

void timer0() interrupt 1
{
    TH0=(65535-50000)/256;
    TL0=(65535-50000)%256;
    aa++;
    if(aa>=20)
    {
        aa=0;
        miao++;
        if(miao==60)
        {
            miao=00;
            fen++;
            if(fen==60)
            {
                fen=00;
                shi++;
                if(shi==24)
                    shi=0;
            }
        }
    }
    if((fen==01)&&(miao==00))
    {   EA=0; }
    if((ringfen==fen)&&(ringshi==shi))
        b_bell=1;
    else
        b_bell=0;
    }
}
```

仿真实验综合设计——案例 2

基于 AT89c52 的跳舞机系统

一、实验目的

跳舞机作为一种音乐节奏类型的游戏，利用玩家的双脚来跟踪屏幕的箭头完成游戏。设计采用芯片 AT89c52 来完成一个简易的跳舞机系统。系统分为显示、按键、声音等模块，能实现简单的跳舞机输出显示及反馈用户输入功能。屏幕按一定节奏随机产生方向箭头，用户根据屏幕显示，通过按键输入方向信号。系统接收并判断输入对错，通过屏幕显示确认画面并发出不同频率声音来反馈用户输入对错，同时对游戏结果记分，完成游戏。

二、实验原理图

1. 应用系统设计

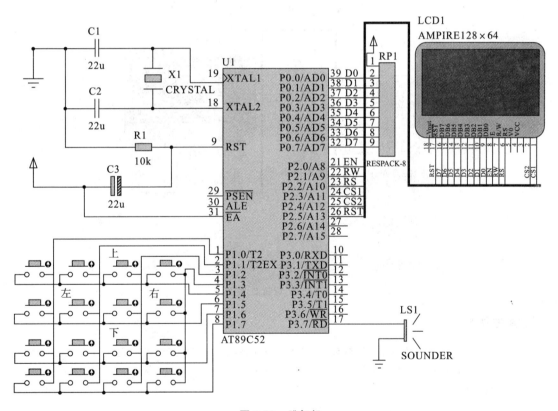

图 3-20 跳舞机

2. 接口设计

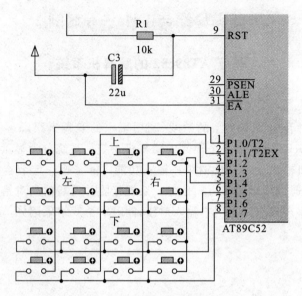

图 3-21 跳舞机键盘接口

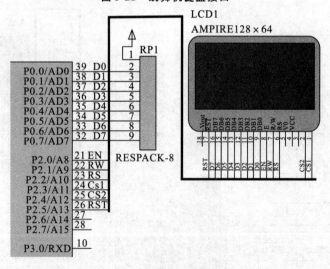

图 3-22 跳舞机显示接口

三、参考实验程序

/ * * * * * * *本程序为不带字库的12864汉字及英文字符的显示程序,还要注意的是:带中文字库和不带中文字库的程序不一样,不可混用 * * * * * * * * * * * * /

```
#include<reg51.h>
#define uchar unsigned char
#define uint unsigned int
//sbit databus=P1;
#define databus P1
//sbit Reset=P3^0;                        //复位
sbit rs=P3^7;                             //指令数据选择
```

```c
sbit e=P3^5;                                    //指令数据控制
sbit cs1=P3^3;                                  //左屏幕选择,低电平有效
sbit cs2=P3^4;                                  //右屏幕选择
sbit wr=P3^6;                                   //读写控制
//sbit busy=P1^7;                               //忙标志
void SendCommand(uchar command);                //写指令
void WriteData(uchar dat);                      //写数据
void LcdDelay(uint time);                       //延时
void SetOnOff(uchar onoff);                     //开关显示
void ClearScreen(uchar screen);                 //清屏
void SetLine(uchar line);                       //置页地址
void SetColum(uchar colum);                     //置列地址
void SetStartLine(uchar startline);             //置显示起始行
void SelectScreen(uchar screen);                //选择屏幕
void Show1616(uchar lin,uchar colum,uchar * address);  //显示一个汉字
void InitLcd();                                 //初始化
void ResetLcd();                                //复位
void Show_english(uchar lin,uchar colum,uchar * address);
const uchar code hzk[]={
/*－－文字:  I  －－*/
/*－－宋体12;  此字体下对应的点阵为:宽×高=8×16   －－*/
0x00,0x08,0x08,0xF8,0x08,0x08,0x00,0x00,0x00,0x20,0x20,0x3F,0x20,0x20,0x00,0x00,

/*－－文字:     －－*/
/*－－宋体12;  此字体下对应的点阵为:宽×高=8×16   －－*/
0x00,0x00,0x00,0x00,0x00,0x00,0x00,0x00,0x00,0x00,0x00,0x00,0x00,0x00,0x00,0x00,
/*－－文字:  c  －－*/
/*－－宋体12;  此字体下对应的点阵为:宽×高=8×16   －－*/
0x00,0x00,0x00,0x80,0x80,0x80,0x00,0x00,0x00,0x0E,0x11,0x20,0x20,0x20,0x11,0x00,
/*－－文字:  a  －－*/
/*－－宋体12;  此字体下对应的点阵为:宽×高=8×16   －－*/
0x00,0x00,0x80,0x80,0x80,0x80,0x00,0x00,0x00,0x19,0x24,0x22,0x22,0x22,0x3F,0x20,

/*－－文字:  n  －－*/
/*－－宋体12;  此字体下对应的点阵为:宽×高=8×16   －－*/
0x80,0x80,0x00,0x80,0x80,0x80,0x00,0x00,0x20,0x3F,0x21,0x00,0x00,0x20,0x3F,0x20,

/*－－文字:     －－*/
/*－－宋体12;  此字体下对应的点阵为:宽×高=8×16   －－*/
0x00,0x00,0x00,0x00,0x00,0x00,0x00,0x00,0x00,0x00,0x00,0x00,0x00,0x00,0x00,

/*－－文字:  m  －－*/
/*－－宋体12;  此字体下对应的点阵为:宽×高=8×16   －－*/
```

0x80,0x80,0x80,0x80,0x80,0x80,0x80,0x00,0x20,0x3F,0x20,0x00,0x3F,0x20,0x00,0x3F,

/*--文字: a --*/
/*--宋体12; 此字体下对应的点阵为:宽×高=8×16   --*/
0x00,0x00,0x80,0x80,0x80,0x80,0x00,0x00,0x00,0x19,0x24,0x22,0x22,0x22,0x3F,0x20,

/*--文字: k --*/
/*--宋体12; 此字体下对应的点阵为:宽×高=8×16   --*/
0x08,0xF8,0x00,0x00,0x80,0x80,0x80,0x00,0x20,0x3F,0x24,0x02,0x2D,0x30,0x20,0x00,

/*--文字: e --*/
/*--宋体12; 此字体下对应的点阵为:宽×高=8×16   --*/
0x00,0x00,0x80,0x80,0x80,0x80,0x00,0x00,0x00,0x1F,0x22,0x22,0x22,0x22,0x13,0x00,

/*--文字: --*/
/*--宋体12; 此字体下对应的点阵为:宽×高=8×16   --*/
0x00,0x00,0x00,0x00,0x00,0x00,0x00,0x00,0x00,0x00,0x00,0x00,0x00,0x00,0x00,0x00,

/*--文字: i --*/
/*--宋体12; 此字体下对应的点阵为:宽×高=8×16   --*/
0x00,0x80,0x98,0x98,0x00,0x00,0x00,0x00,0x00,0x20,0x20,0x3F,0x20,0x20,0x00,0x00,

/*--文字: t --*/
/*--宋体12; 此字体下对应的点阵为:宽×高=8×16   --*/
0x00,0x80,0x80,0xE0,0x80,0x80,0x00,0x00,0x00,0x00,0x00,0x1F,0x20,0x20,0x00,0x00,

/*--文字: ! --*/
/*--宋体12; 此字体下对应的点阵为:宽×高=16×16   --*/
0x00,0x00,0x00,0x00,0xF0,0x00,0x00,0x00,0x00,0x00,0x00,0x00,0x00,0x00,0x00,0x00,
0x00,0x00,0x00,0x5F,0x00,0x00,0x00,0x00,0x00,0x00,0x00,0x00,0x00,0x00,0x00,0x00,

/*--文字: 我 --*/
/*--楷体_GB231212; 此字体下对应的点阵为:宽×高=16×16   --*/
0x00,0x00,0x80,0x90,0xF0,0x48,0x40,0x7F,0xC0,0x20,0x24,0xA8,0x00,0x00,0x00,0x00,
0x08,0x08,0x04,0x14,0x3F,0x02,0x09,0x08,0x05,0x06,0x09,0x10,0x20,0x78,0x00,0x00,

/*--文字: 的 --*/
/*--楷体_GB231212; 此字体下对应的点阵为:宽×高=16×16   --*/
0x00,0xC0,0x60,0x50,0x2C,0xE0,0x80,0x40,0xA0,0x38,0x26,0x10,0xF0,0x00,0x00,0x00,
0x00,0x07,0x19,0x09,0x08,0x1F,0x00,0x00,0x00,0x03,0x10,0x20,0x1F,0x00,0x00,0x00,

/*--文字: 未 --*/
/*--楷体_GB231212; 此字体下对应的点阵为:宽×高=16×16   --*/
0x00,0x00,0x80,0x80,0x90,0x90,0xFF,0xC8,0x48,0x48,0x40,0x40,0x00,0x00,0x00,0x00,
0x10,0x10,0x08,0x04,0x02,0x01,0x7F,0x00,0x01,0x02,0x04,0x08,0x18,0x10,0x10,0x00,

/*——文字： 来 ——*/
/*——楷体_GB231212； 此字体下对应的点阵为:宽×高=16×16 ——*/
0x00,0x80,0x80,0xA8,0xC8,0x88,0xFF,0x84,0x64,0x54,0x40,0x40,0x00,0x00,0x00,0x00,
0x00,0x10,0x10,0x08,0x04,0x02,0x7F,0x01,0x02,0x04,0x0C,0x08,0x08,0x08,0x08,0x00,

/*——文字： 不 ——*/
/*——楷体_GB231212； 此字体下对应的点阵为:宽×高=16×16 ——*/
0x00,0x00,0x08,0x08,0x08,0x88,0x48,0xE4,0x14,0x8C,0x84,0x04,0x04,0x04,0x00,0x00,
0x00,0x04,0x04,0x02,0x01,0x00,0x00,0x3F,0x00,0x00,0x00,0x01,0x03,0x06,0x00,0x00,

/*——文字： 是 ——*/
/*——楷体_GB231212； 此字体下对应的点阵为:宽×高=16×16 ——*/
0x00,0x00,0x80,0x80,0x82,0x9E,0xAA,0xAA,0xA1,0x5D,0x43,0x40,0x00,0x00,0x00,0x00,
0x20,0x20,0x10,0x08,0x06,0x04,0x08,0x1F,0x12,0x22,0x22,0x20,0x20,0x20,0x20,0x00,

/*——文字： 梦 ——*/
/*——楷体_GB231212； 此字体下对应的点阵为:宽×高=16×16 ——*/
0x00,0x90,0x50,0x30,0xFE,0x28,0x48,0x28,0x18,0xFF,0x14,0x24,0x24,0x40,0x40,0x00,
0x00,0x00,0x40,0x48,0x44,0x26,0x2B,0x12,0x0A,0x06,0x00,0x00,0x00,0x00,0x00,0x00,

/*——文字： ! ——*/
/*——楷体_GB231212； 此字体下对应的点阵为:宽×高=16×16 ——*/
0x00,0x00,0x00,0xFC,0xFC,0x00,0x00,0x00,0x00,0x00,0x00,0x00,0x00,0x00,0x00,0x00,
0x00,0x00,0x00,0x19,0x19,0x00,0x00,0x00,0x00,0x00,0x00,0x00,0x00,0x00,0x00,0x00,
};
void main()
{
    uchar i,line,colum,j ;
    uchar *address ;
    InitLcd();
    while(1)
    {
    /*显示第一行*/
    /*******************下面这段程序用来卷页******************/
    /*line=0;
        for(j=0;j<4;j++)
        {
            ClearScreen(2);   //清屏
            line=line+1;
            colum=0;
            address=hzk;
            SetOnOff(1);

```c
            for(i=0;i<14;i++)
            {
                if(i<8)
                {
                    SelectScreen(0);
                    Show_english(line,colum,address);
                    address+=16;
                    colum+=8;
                }
                else if(i>=8)
                {
                    if(i<13)
                    {
                        SelectScreen(1);
                        Show_english(line,colum,address);
                        address+=16;
                        colum+=8;
                    }
                    else
                    {
                        Show1616(line,colum,address);
                        address+=32;
                        colum+=16;
                    }

                }
            }
            for(i=0;i<50;i++)               //延时
                LcdDelay(3000);
}   */
    line=1;                             //从第2页(第9行)开始显示
    colum=0;                            //从第一列开始显示
    address=hzk;                        //给地址指针赋初值
    SetOnOff(1);                        //显示开,注意:如果这里设置显示关,显示会出现错误
    for(i=0;i<14;i++)                   //设置要显示的字符个数
    {
        if(i<8)    //i<8时,在左半屏显示(因为每半屏最多只能显示8个英文字符即4个汉字)
        {
            SelectScreen(0);                //选择左屏
            Show_english(line,colum,address);   //显示一个英文字符
            address+=16;                    //每个英文字符需要16个十六进制数表示
            colum+=8;                       //每个英文字符占8列
        }
```

```c
        else if(i>=8)                          //当 i>8 时(当然最多只能是 16)在右屏显示
        {
            if(i<13)    //第一行前 13 个字符为英文字符,最后一个字符为中文字符,英文字符和中文
                        //字符必须分开显示
            {
                SelectScreen(1);
                Show_english(line,colum,address);
                address+=16;
                colum+=8;
            }
            else                                //显示中文字符"!"
            {
                Show1616(line,colum,address);    //显示一个汉字
                address+=32;                     //每个汉字要用 32 个十六进制表示
                colum+=16;                       //每个汉字占 16 列
            }
        }
    }
/*显示第二行*/
/*****显示原理与第一行完全相同,这里不再赘述****/
line=4;
colum=1;
SetOnOff(1);
for(i=0;i<8;i++)
{
    if(i<4)
    {
        SelectScreen(0);
        Show1616(line,colum,address);
        address+=32;
        colum+=16;
    }
    else
    {
        SelectScreen(1);
        Show1616(line,colum,address);
        address+=32;
        colum+=16;
    }
}
SetOnOff(1);
for(i=0;i<50;i++)                              //延时
LcdDelay(30000);
```

```c
        if(colum>63)
            colum=0;
    }
}
/*********************延时函数************************/
void LcdDelay(uint time)
{
    while(time--);
}
/*********************写指令************************/
void SendCommand(uchar command)
{
    e=1;
    wr=0;
    rs=0;
    databus=command;
    e=0;
}
/*********************写数据************************/
void WriteData(uchar dat)
{
    e=1;
    wr=0;
    rs=1;
    databus=dat;
    e=0;
}
/*********************显示开/关************************/
void SetOnOff(uchar onoff)
{
    if(onoff==1)
    {
        SendCommand(0x3f);
    }
    else
    {
        SendCommand(0x3e);
    }
}
/*********************选择页************************/
void SetLine(uchar line)              //12864 总共有 8 页(0~7),每页有 8 行
{
    line=line&0x07;                   //只取后三位 xxxx x111,这 3 个是要改变位置的数据
```

```c
    line=line|0xb8;                    //页设置的固定格式
    SendCommand(line);
}
/************************选择列*********************/
void SetColum(uchar colum)             //12864每半屏有64列(0~63),分为左右2屏
{
    colum=colum&0x3f;                  //xx11 1111,这个是要改变Y位置的数据
    colum=colum|0x40;                  //固定格式
    SendCommand(colum);
}
/***********************选择起始行*********************/
void SetStartLine(uchar startline)
{
    startline=startline&0x3f;          //xx11 1111,这个是要改变x位置的数据
    startline=startline|0xc0;          //11xxxxxx,是起始行设置的固定指令
    SendCommand(startline);
}
/*********选择左右屏 0:左屏,1:右屏,2:全屏***************/
void SelectScreen(uchar screen)
{
    switch(screen)
    {
        case 0:
            cs1=0;
            LcdDelay(2);
            cs2=1;
            LcdDelay(2);
            break;
        case 1:
            cs1=1;
            LcdDelay(2);
            cs2=0;
            LcdDelay(2);
            break;
        case 2:
            cs1=0;
            LcdDelay(2);
            cs2=0;
            LcdDelay(2);
            break;
    }
}
/***********************显示一个汉字*********************/
```

```c
void Show1616(uchar lin,uchar colum,uchar * address)
{
    uchar i;
    SetLine(lin);
    SetColum(colum);
    for(i=0;i<16;i++)
    {
        WriteData(*address);
        address++;
    }
    SetLine(lin+1);
    SetColum(colum);
    for(i=0;i<16;i++)
    {
        WriteData(*address);
        address++;
    }
}
/*********************显示一个英文字符***********************/
void Show_english(uchar lin,uchar colum,uchar * address)
{
    uchar i;
    SetLine(lin);
    SetColum(colum);
    for(i=0;i<8;i++)
    {
        WriteData(*address);
        address++;
    }
    SetLine(lin+1);
    SetColum(colum);
    for(i=0;i<8;i++)
    {
        WriteData(*address);
        address++;
    }
}
/***********************清屏***************************/
void ClearScreen(uchar screen)
{
    uchar i,j;
    SelectScreen(screen);
    for(i=0;i<8;i++)
```

```
        {
            SetLine(i);
            SetColum(0);
            for(j=0;j<64;j++)
            {
                WriteData(0);
            }
        }
}
/******************12864初始化*********************/
void InitLcd()
{
//    ResetLcd();
    SetOnOff(0);                        //显示关
    ClearScreen(2);                     //清屏
    SetLine(0);                         //页设置
    SetColum(0);                        //列设置
    SetStartLine(0);                    //设置起始页
    SetOnOff(1);                        //显示开
}
```

实验 8 PWM 实验

一、实验目的

1. 了解 PWM 电压转换原理。
2. 掌握单片机控制的 PWM 电压转换。

二、实验设备

STAR 系列实验仪一套、PC 机一台。

三、实验内容

1. PWM 电压转换原理
①将一定频率的输入信号转换为直流电。
②通过调节输入信号占空比调节输出的直流电电压,输出电压随着占空比增大而减小。

2. 实验过程
①输入 15kHz 左右的方波,经 LM358 进行 PWM 电压转换,输出直流电,驱动直流电机。
②通过按键调整占空比来改变 PWM 输出电压,直流电机的转速会随之变化。

四、实验原理图

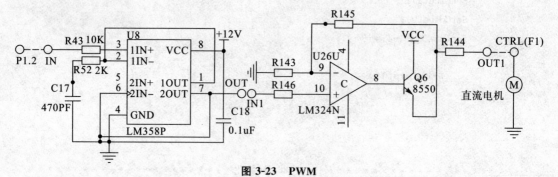

图 3-23　PWM

五、实验步骤

1. 连线说明：

E2 区:IN	——	A3 区:P1.2,方波输入
E2 区:OUT	——	E2 区:IN1
E2 区:OUT1	——	F1 区:CTRL,直流电机电源输入
A3 区:JP51	——	G6 区:JP74

2. 通过 G6 区的 1、2 键调整占空比来改变 PWM 输出电压,直流电机的转速会随之变化:1 号键减少占空比;2 号键增加占空比。

六、演示程序

```
            INBITP1.2                       ;PWM 方波输入
PWM_LOW     DATA      30H                   ;低电平时间
PWM_HIGH    DATA      31H                   ;高电平时间,控制频率在 15kHz 左右
periods     EQU       0E0H                  ;周期 64us
            ORG       0000H
            LJMP      START
            ORG       000BH
            LJMP      iTIMER0
            ORG       0100H
START:      MOV       SP,#60H
            MOV       PWM_LOW,#periods
            MOV       PWM_HIGH,#periods
            MOV       TH0,#periods
            MOV       TL0,#periods
            MOV       TMOD,#02H
            SETB      EA
            SETB      ET0
            SETB      TR0
START1:     ACALL     ScanKey
            JNZ       Key1
```

Key0:	MOV	A,PWM_HIGH	;增加占空比
	CJNE	A,#0FBH,Key0_1	
	SJMP	START1	;大于这个值,对定时中断已反应不过来
Key0_1:	INC	PWM_HIGH	
	DEC	PWM_LOW	
	SJMP	START1	
Key1:	MOV	A,PWM_LOW	;减少占空比
	CJNE	A,#0FBH,Key1_1	
	SJMP	START1	;大于这个值,对定时中断已反应不过来
Key1_1:	INC	PWM_LOW	
	DEC	PWM_HIGH	
	SJMP	START1	
iTIMER0:	JBC	IN,iTIMER0_1	
	MOV	TL0,PWM_HIGH	
	SETB	IN	
	RETI		
iTIMER0_1:	MOV	TL0,PWM_LOW	
	NOP		
	RETI		
ScanKey:	JNB	P1.0,ScanKey1	;键扫描
	JB	P1.1,ScanKey	
ScanKey1:	ACALL	Delay20ms	;消抖动
	ACALL	Delay20ms	
	JNB	P1.0,ScanKey2	
	JB	P1.1,ScanKey	
	MOV	A,#1	
	SJMP	ScanKey3	
ScanKey2:	CLR	A	
ScanKey3:	JNB	P1.0,$	
	JNB	P1.1,$	
	RET		
Delay20ms:	MOV	R6,#10	
Delay1:	MOV	R7,#100	
	DJNZ	R7,$	
	DJNZ	R6,Delay1	
	RET		
	END		

七、实验扩展及思考

改变 PWM 的输入频率,使用示波器观看 LM358 的输出,由此加深对 PWM 电压转换的了解。

实验 9 8255 控制交通灯实验

一、实验目的

1. 了解 8255 芯片的工作原理,熟悉其初始化编程方法以及输入、输出程序设计技巧。学会使用 8255 并行接口芯片实现各种控制功能,如本实验(控制交通灯)等。

2. 熟悉 8255 内部结构与单片机的接口逻辑,熟悉 8255 芯片的 3 种工作方式以及控制字格式。

3. 认真预习本节实验内容,尝试自行编写程序,填写实验报告。

二、实验设备

STAR 系列实验仪一套、PC 机一台。

三、实验内容

1. 编写程序:使用 8255 的 PA0..2、PA5..7 控制 LED 指示灯,实现交通灯功能。
2. 连接线路验证 8255 的功能,熟悉它的使用方法。

四、实验原理图

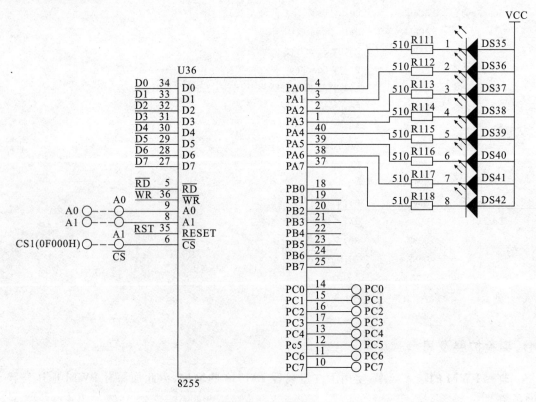

图 3-24 8255 接口

五、实验步骤

1. 连线说明：

B4 区:CS、A0、A1	——	A3 区:CS1、A0、A1
B4 区:JP56(PA 口)	——	G6 区:JP65

2. 观察实验结果，是否能看到模拟的交通灯控制过程。

六、演示程序

```
xdata unsigned char com_address _at_ 0xf003;
xdata unsigned char pa _at_ 0xf000;
const unsigned char Led_const[]={0x7e,    //东西绿灯,南北红灯
                                 0xfe,    //东西绿灯闪烁,南北红灯
                                 0xbe,    //东西黄灯亮,南北红灯
                                 0xdb,    //东西红灯,南北绿灯
                                 0xdf,    //东西红灯,南北绿灯闪烁
                                 0xdd};   //东西红灯,南北黄灯亮

void delay500ms()
{
    unsigned int i;
    for (i=0; i< 0xffff; i++)
    {;};
}

void delay3s()
{
    unsigned int i=6;
    while (i--)
        delay500ms();
}

void delay5s()
{
    unsigned int i=10;
    while (i--)
        delay500ms();
}

main()
{
    unsigned char j;
```

```c
        com_address=0x80;          //PA、PB、PC 为基本输出模式
        pa=0xff;                   //灯全熄灭
        while (1)
        {
            pa=Led_const[0];
            delay5s();             //东西绿灯,南北红灯
            j=6;
            while(j--)
            {
                pa=Led_const[1];   //东西绿灯闪烁,南北红灯
                delay500ms();
                pa=Led_const[0];
                delay500ms();
            }
            pa=Led_const[2];       //东西黄灯亮,南北红灯
            delay3s();
            pa=Led_const[3];       //东西红灯,南北绿灯
            delay5s();
            j=6;
            while (j--)
            {
                pa=Led_const[4];   //东西红灯,南北绿灯闪烁
                delay500ms();
                pa=Led_const[3];
                delay500ms();
            }
            pa=Led_const[5];       //东西红灯,南北黄灯亮
            delay3s();
        }
    }
```

七、实验扩展及思考

如何对 8255 的 PC 口进行位操作？

实验 10　8155 输入、输出、SRAM 实验

一、实验目的

1. 了解 8155 的内部资源与结构；了解 8155 与单片机的接口逻辑；熟悉对 8155 的初始化编程、输入和输出程序的设计方法、8155 定时器/计数器的使用方法。

2. 认真预习，做好实验前的准备工作，填写实验报告。

二、实验设备

STAR 系列实验仪一套、PC 机一台。

三、实验内容

1. 编写程序：从 8155 的 PA 口将 G6 区的 8 位开关读入，写入 8155 的内部 RAM，再读出后，写入 PB 口，显示于 LED 指示灯上。

2. 连接线路，验证 8155 的功能，熟悉它的使用方法。

四、实验原理图

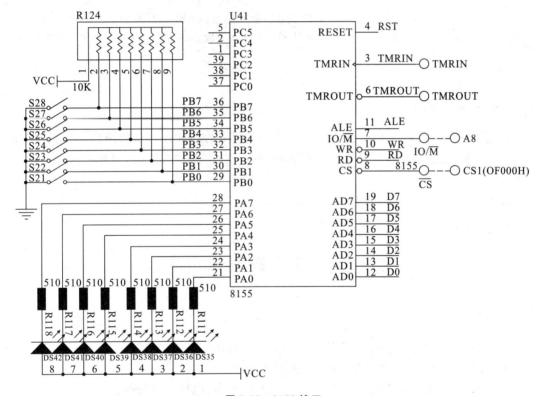

图 3-25　8155 接口

五、实验步骤

1. 连线说明：

B4 区：CS、IO/M	——	A3 区：CS1、A8
B4 区：JP76（PA 口）	——	G6 区：JP65
B4 区：JP75（PB 口）	——	G6 区：JP80

2. 测试实验结果：G6 区的开关状态反应在 G6 区的 LED 指示灯上。

六、演示程序

```
    xdata unsigned char COM_8155 _at_ 0xf100;    //命令字/状态寄存器
    xdata unsigned char PA _at_ 0xf101;          //PA 口地址
    xdata unsigned char PB _at_ 0xf102;          //PB 口地址
    xdata unsigned char RAM_8155 _at_ 0xf000;    //8155 内部 RAM 00 单元地址
    main()
    {
        COM_8155=0x1;                //PA 为基本输出,PB 为基本输入
        while (1)
        {
            RAM_8155=PB;             //从 PB 口获得输入值(拨码盘输入,存入 8155 内部 RAM 里)
            PA=RAM_8155;             //重新从 8155 相同地址取数,输出送显示(8 个发光二极管)
        }
    }
```

七、实验扩展及思考

1. 例子程序中只展示了 8155 的输入输出和读写数据 RAM 的功能,8155 还有定时器/计数器的功能,有兴趣读者可以自己编写程序,通过 8155 来实现定时,当作定时器用时,如何接线?

2. 若对 PC 口进行位操作,应该如何编写程序?

3. 如何使用 8155 实现键盘扫描和 LED 显示?

实验 11　8279 键盘显示实验

一、实验目的

1. 了解 8279 的内部结构、工作原理;了解 8279 与单片机的接口逻辑;掌握对 8279 的编程方法,掌握使用 8279 扩展键盘、显示器的方法。

2. 认真预习,做好实验前的准备工作,自行编写程序,填写实验报告。

二、实验设备

STAR 系列实验仪一套、PC 机一台。

三、实验内容

1. 编写程序:利用 8279 实现对 G5 区的键盘扫描,将键号显示于 8 位数码管上。

2. 按图连线,运行程序,观察实验结果,能熟练运用 8279 扩展显示器和键盘。

四、实验原理图

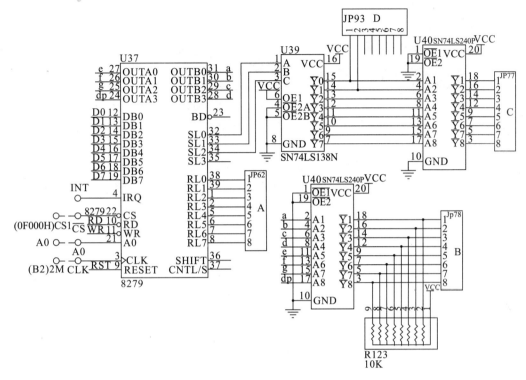

图 3-26　8279 接口

五、实验步骤

1. 连线说明：

E5 区：CS、A0	——	A3 区：CS5、A0
E5 区：CLK	——	B2 区：2M
E5 区：A、B、C、D	——	G5 区：A、B、C、D

2. 运行程序，观察实验结果（任意按下 G5 区 4×4 键盘几个键，它上面的 8 个 LED 显示器会将按键的编码从左至右依次显示出来），可依此验证对 8279 芯片操作的正确性。

六、演示程序

```
xdata unsigned char CMD_8279 _at_ 0xbf01;   //8279 命令字、状态字地址
xdata unsigned char DATA_8279 _at_ 0xbf00;  //8279 读写数据口的地址

const unsigned char LED_TAB[]={0xc0,0xF9,0xA4,0xB0,0x99,0x92,0x82,0xF8,0x80,0x90,0x88,
                               0x83,0xC6,0xA1,0x86,0x8E};
void clear()
{
    CMD_8279=0xde;                          //清除命令
```

```c
        while (CMD_8279 & 0x80);          //显示 RAM 清除完毕吗
}


//8279 初始化
void init8279()
{
    CMD_8279=0x34;                        //可编程时钟设置,设置分频系数(20 分频)
    CMD_8279=0;                           //8×8 字符显示,左边输入,外部译码键扫描方式
    clear();                              //清显示
    CMD_8279=0x90;                        //从第一个数码管开始移位显示
}


void init8279_1()
{
    clear();                              //清显示
    CMD_8279=0x90;                        //从第一个数码管开始移位显示
}


bit scan_key(unsigned char * pt)
{
    if (CMD_8279 & 7)
    {                                     //有键按下
        CMD_8279=0x40;                    //读 FIFO RAM
        * pt=DATA_8279;
        return 1;
    }
    else
        return 0;                         //无键按下,返回标志 0
}


unsigned char key_num(unsigned char i)
{
    return (i & 0x3f);
}


void write_data(unsigned char i)
{
    DATA_8279=i;
}


main()
{
    unsigned char key;
```

```
    unsigned char keyCount=0;
    init8279();                         //初始化子程序
    while(1)
    {
        while(!scan_key(&key));         //键扫描
        keyCount++;
        if(keyCount==8)
        {
            keyCount=0;
            init8279_1();               //8个数码块全有字符显示后,再按键,清除显示
        }
        write_data(LED_TAB[key_num(key)]);
        //键值转换为键号,根据字形码表,取字形,送显示
    }
}
```

七、实验扩展及思考

重新编写软件实验 2,自己编写键扫描、显示程序。

实验 12　并行 DA 实验

一、实验目的

了解数模转换的原理;了解 0832 与单片机的接口逻辑;掌握使用 DAC0832 进行数模转换。

二、实验设备

STAR 系列实验仪一套、PC 机一台、示波器一台。

三、实验内容

1. 编写程序:用 0832 输出正弦波。
2. 按图连线,运行程序,使用示波器观察实验结果。

四、实验原理图

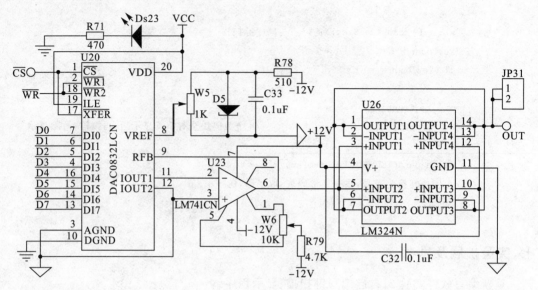

图 3-27　并行接口

五、实验步骤

1. 连线说明：

| F3 区:CS | —— | A3 区:CS1 |

2. 运行程序,示波器的探头接 F3 区的 OUT,观察实验结果,是否产生正弦波。

六、演示程序

```
const unsigned char TAB_1[]=
{0x7F,0x8B,0x96,0xA1,0xAB,0xB6,0xC0,0xC9,0xD2,0xDA,0xE2,0xE8,0xEE,
0xF4,0xF8,0xFB,0xFE,0xFF,0xFF,0xFF,0xFE,0xFB,0xF8,0xF4,0xEE,0xE8,
0xE2,0xDA,0xD2,0xC9,0xC0,0xB6,0xAB,0xA1,0x96,0x8B,0x7F,0x74,0x69,
0x5E,0x54,0x49,0x40,0x36,0x2D,0x25,0x1D,0x17,0x11,0x0B,0x7,0x4,0x2,
0x0,0x0,0x0,0x2,0x4,0x7,0x0B,0x11,0x17,0x1D,0x25,0x2D,0x36,0x40,0x49,0x54,0x5E,0x69,0x74};
xdata unsigned char Addr_0832 _at_ 0xff00;  //0832 输出口地址

void delay()
{
    unsigned char i=0x50;
    while(i--)
    {;}
}
```

```
main()
{
    char i;
    while(1)
    {
        for(i=0; i<72; i++)
        {
            Addr_0832=TAB_1[i];
            delay();
        }
    }
}
```

实验 13　并行 AD 实验（数字电压表实验）

一、实验目的

1. 了解几种类型 AD 转换的原理；掌握使用 ADC0809 进行模数转换。
2. 认真预习实验内容，做好准备工作，完成实验报告。

二、实验设备

STAR 系列实验仪一套、PC 机一台、万用表一个。

三、实验内容

1. ADC0809（G4 区）。
①模数转换器，8 位精度，8 路转换通道，并行输出。
②转换时间 100us，转换电压范围 0~5V。
2. 编写程序：制作一个电压表，测量 0~5V，结果显示于数码管上。

四、实验原理图

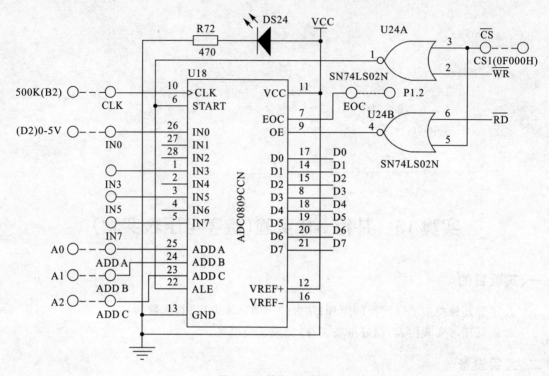

图 3-28 并行 AD 接口

五、实验步骤

1. 连线说明:

G4 区:CS、ADDA、ADDB、ADDC	——	A3 区:CS1、A0、A1、A2(选择通道)
G4 区:EOC(转换结束标志)	——	A3 区:P1.2
G4 区:CLK	——	B2 区:500K
G4 区:IN0	——	D2 区:0~5V
E5 区:CLK	——	B2 区:2M
E5 区:CS	——	A3 区:CS5
E5 区:A0	——	A3 区:A0
E5 区:A、B、C、D	——	G5 区:A、B、C、D

2. 调节 0~5V 电位器(D2 区)输出电压,显示在 LED 上,第 4、5 位显示十六进制数据,第 0、1、2 位显示十进制数据。用万用表验证 AD 转换的结果。

六、演示程序

```
extern void Display8();
#include "reg52.h"
#include "stdio.h"
xdata unsigned char Addr_0809 _at_ 0xf000;
```

```c
data unsigned char buffer[8];              //8个字节的显示缓冲区
sbit EOC_0809=P1^2;
data unsigned char R0 _at_ 0x0;
unsigned char AD0809()
{
    Addr_0809=0;                           //启动 AD 转换
    while (!EOC_0809);                     //是否转换完成
    return Addr_0809;                      //读转换结果
}

void delay()
{
    int i;
    for (i=0; i < 0x3fff; i++)
    {;}
}

main()
{
    unsigned char adResult;
    float temp;
    while (1)
    {
        adResult=AD0809();
        temp=adResult;
        temp /=51;
        sprintf(buffer+4,"%1.2f",temp);
        buffer[2]=(buffer[4]-0x30+0x80);   //加上小数点
        buffer[1]=buffer[6]-0x30;
        buffer[0]=buffer[7]-0x30;
        buffer[4]=adResult & 0xf;
        buffer[5]=((adResult & 0xf0) >> 4);
        buffer[3]=0x10;
        buffer[6]=0x10;
        buffer[7]=0x10;                    //消隐
        R0=buffer;
        Display8();                        //库函数 Display8 需要使用 R0 指明显示缓冲区
        delay();
    }
}
```

七、实验扩展及思考

如何实现多路模拟量的数据采集、显示?

实验 14　红外通信实验

一、实验目的

1. 理解红外通讯原理。
2. 掌握红外通讯。

二、实验设备

STAR 系列实验仪一套、PC 机一台。

三、实验内容

1. 红外通讯原理

当红外接收器收到 38kHZ 频率的信号,输出电平会由 1 变为 0,一旦没有此频率信号,输出电平会由 0 变为 1。因此,红外发射头控制通断发射 38kHZ 信号,就可以将数据发送出来。

2. 实验过程

①使用红外发送管和接收器进行数据自发自收。
②根据接收到的数据点亮 P1 口的 8 个发光管,会看到发光管不断变化。

四、实验原理图

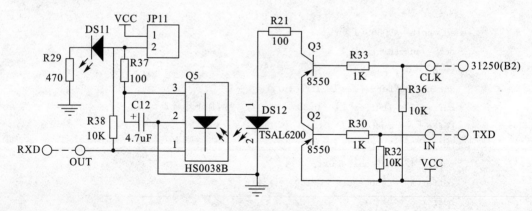

图 3-29　红外接口

五、实验步骤

1. 连线说明:

G2 区:IN	A3 区:TxD
G2 区:OUT	A3 区:RXD
G2 区:CLK	B2 区:31250
A3 区:JP51(P1)	G6 区:JP65

2. 调试该程序时,使用较厚的白纸挡住红外发射管红外信号,使它反射到接收头。

说明:一般红外接收模块的解调频率为 38kHz,当它接收到 38kHz 左右的红外信号时将输出低电平,但连续输出低电平的时间是有限制的(如:100ms),也就是说输出低电平的宽度是有限制的。

3. 发送数据,并接收,根据接收到的数据点亮 8 个发光管,程序运行之后,会看到 8 个发光管(G6 区)在闪烁,从第 8 个(最右边)向第 1 个逐一点亮过去。本实验通过红外通讯发送、接收数据,发送的数据从 00H 开始+1,接收到该数据后用来点亮 8 个发光管。亮-1,熄-0。

六、演示程序

```
;初始化
#include "reg52.h"
void Infrared_INIT()                //初始化
{
    P1=0xff;                        //令发光管灭
    TMOD=0x20;                      //定时器工作方式 2,设波特率 2400
    TH1=0xf4;
    TL1=0xf4;
    TR1=1;                          //选通定时器 1,定时器开始工作
    SCON=0x50;                      //串口工作方式 1,开允许接收
}

//延时 0.1s
void Delay_01ms()
{
    unsigned char i=50;
    while(i--);
}


//延时程序
void Delay()
{
    unsigned int i;
    for(i=0;i<0xffff;i++)
    {;}
}

//红外通讯数据自收自发子程序
unsigned char Send_Receive(unsigned char i)
{
    unsigned char count=0x60;       //检测接收标志最大次数
    TI=0;
```

```c
        SBUF=i;
        while(count--)
        {
            if(RI)
            {
                RI=0;
                return SBUF;
            }
            Delay_01ms();              //每隔0.1ms检测一次接收标志
        }

        return 0;                      //超时
}

//红外通讯
void Infrared_Test()
{
    static unsigned char i=1;
    unsigned char j;
    j=Send_Receive(i);                 //红外通讯
    P1=~j;                             //根据收接到的数据点亮8个红色发光管
    i++;                               //发送数据逐步递增
}
main()
{
    Infrared_INIT();                   //红外通讯初始化
    while(1)
    {
        Infrared_Test();               //调用自收自发红外通讯子程序
        Delay();                       //延时
    }
}
```

七、实验扩展及思考

了解日常所用的家电红外遥控器是如何工作的,结合按键模拟4路红外遥控器,编写程序遥控发光管或电机转动快慢。

实验15 X5045 串行 EEPROM 读写实验

一、实验目的

1.了解 SPI 总线;掌握 SPI 总线的数据读写操作;掌握对 X5045 单字节读写、页写模式和连续读取模式的操作。

2. 学会建立项目,使用多个模块文件完成一个功能,便于移植。

二、实验设备

STAR 系列实验仪一套、PC 机一台。

三、实验内容

1. X5045(C4 区)

①内置 4K bit 三线串行接口 EEPROM,支持 16 字节页写模式和连续读取功能。

②内置看门狗和低电压检测功能。

2. 实验过程

①写满 X5045 内部整个 4K bit 串行 EEPROM,起始写入地址为 00H,之后地址与数据都以 +1 递增,直到写满整个 EEPROM。

②检验写入数据是否正确并显示结果,正确:点亮 G6 区的发光管,错误:熄灭发光管。

四、实验原理图

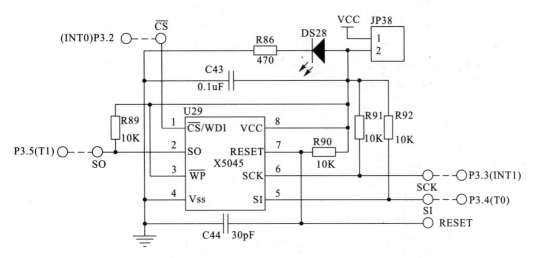

图 3-30 串行 EPROM 接口

五、实验步骤

1. 主机连线说明:

C4 区:CS、SCK	——	A3 区:P3.2(INT0)、P3.3(INT1)
C4 区:SI、SO	——	A3 区:P3.4(T0)、P3.5(T1)
G6 区:JP65	——	A3 区:JP51(P1 口)

2. 运行程序,正确:点亮 G6 区的发光管,错误:熄灭发光管。

六、演示程序

```c
//;************************************************************
//;X5045 操作子程序
//;X5045 内置 4K bits 三线串行接口 EEPROM,支持 16 字节页写模式和
//;连续读取功能。内置看门狗和低电压检测功能
//;************************************************************
#include "reg52.h"
#include <intrins.h>

sbit CS=P3^2;                    //片选,低有效
sbit SCLK=P3^3;                  //时钟输入
sbit SI=P3^4;                    //数据输入
sbit SO=P3^5;                    //数据输出

//写数据
#define X_WMidXTRAN
//读数据
#define X_RMidXRECE

//发送8个位的数据/命令
void XTRAN(unsigned char _data)
{
    unsigned char i=0x80;
    while (i)
    {
        SCLK=0;                  //上升沿写入
        if (_data & i)
            SI=1;
        else
            SI=0;
        i>>=1;
        SCLK=1;
    }
}

//接收8个位的数据/命令
unsigned char XRECE()
{
    unsigned char i=0;
    unsigned char j=0x80;
    while (j)
    {
```

```
        SCLK=1;                              //下降沿读出
        _nop_();
        SCLK=0;
        if (SO)
            i|=j;
        j>>=1;
    }
    return i;
}

//读状态
unsigned char X_RSDR()
{
    SCLK=0;                                  //读状态,用于查忙
    CS=0;
    XTRAN(5);
    ACC=XRECE();
    CS=1;
    return ACC;
}

//允许写操作
void X_WREN()
{
    SCLK=0;
    CS=0;
    XTRAN(6);                                //允许写指令06H
    CS=1;
}

//写状态,控制看门狗,存储内容的保护范围
void X_WRSR(unsigned char _data)
{
    X_WREN();                                //必须先开启允许写
    SCLK=0;
    CS=0;
    XTRAN(1);                                //写状态寄存器命令01H
    XTRAN(_data);
    CS=1;
    while (X_RSDR() & 1)
    {;}
}
```

```c
//禁止写操作
void X_WRDI()
{
    SCLK=0;
    CS=0;
    XTRAN(4);                          //禁止写指令 04H
    CS=1;
}

//开始写
void X_WBeg(bit bLowAddress,unsigned char address)
{
    X_WREN();                          //每次要写入数据,都必须开允许写操作
    SCLK=0;
    CS=0;
    if (bLowAddress)
        XTRAN(2);                      //写低 256 个字节数据
    else
        XTRAN(0xa);                    //写高 256 个字节数据
    XTRAN(address);
}

//结束写
void X_WEnd()
{
    CS=1;                              //写数据结束后,关闭允许写
    while (X_RSDR() & 1)
    {;}
}

//开始读
void X_RBeg(bit bLowAddress, unsigned char address)
{
    SCLK=0;
    CS=0;
    if (bLowAddress)
        XTRAN(3);                      //读低 2K bits 数据
    else
        XTRAN(0xb);                    //读高 2K bits 数据
    XTRAN(address);
}

//结束读
```

```c
void X_REnd()
{
    CS=1;
}
```

//测试 X5045 测试
//写数据,X5045 内的串行 EEPROM 分为高、低各 256 个字节,要分别写入

```c
#define WLOOP1 0x10                 //写内循环次数
#define WLOOP2 0x20                 //写外循环次数
#define RLOOP1 0x10                 //读内循环次数
#define RLOOP2 0x20                 //读外循环次数
#define T_Data 0xFE                 //起始数据 FEH

void T_X5045_WR()
{
    bit bLowAddress=1;
    unsigned char _data=T_Data;     //起始数据,之后数据值会不断增加
    unsigned char address=0;        //串行 EEPROM 地址起始地址
    char i,j;
    X_WRSR(0x30);                   //写状态寄存器
    for (i=0; i< WLOOP2; i++)       //外循环次数 20H
    {
        if ( i>=0x11)
            bLowAddress=0;
        X_WBeg(bLowAddress , address);
        for (j=0; j< WLOOP1; j++)
            X_WMid(_data++);        //写操作
        X_WEnd();                   //结束本次写操作
        address+=WLOOP1;
    }
}
```

//读数据,可以一次连续地读出高、低 256 字节串行 EEPROM 里的数据,并检验
```c
bit T_X5045_RD()
{
    bit bLowAddress=1;
    unsigned char _data=T_Data;     //起始数据,之后数据值会不断增加
    unsigned char address=0;        //串行 EEPROM 地址起始地址
    char i,j;
    unsigned char rdata;
    X_WRSR(0x30);                   //写状态寄存器
    X_RBeg(0, address);
```

```c
        for (i=0; i< WLOOP2; i++)              //外循环次数 20H
        {
            for (j=0; j< WLOOP1; j++)
            {
                rdata=X_RMid();                 //写操作
                if (rdata!=_data)
                {
                    X_REnd();
                    return 0;
                }
                _data++;
            }
        }
        X_REnd();                               //读数据结束
        return 1;
}
//主程序 MAIN.C
#include "reg52.h"
extern bit T_X5045_RD();
extern void T_X5045_WR();
//X5045 的测试实验
void X5045TEST()
{
    T_X5045_WR();           //调用 X5045 数据写入程序,该程序在 X5045.C 中
    if (T_X5045_RD())       //调用 X5045 数据检验程序,读出 X5045 内的数据与先前的写
                            //入数据相比较,看是否正确,该程序在 X5045.C 中
        P1=0;               //测试正确,点亮 8 个发光管
    else
        P1=0xff;            //测试错误,熄灭 8 个发光管
}

main()
{
    X5045TEST();            //X5045 测试实验
    while (1)
    {;}
}
```

七、实验扩展及思考

上面实验采用连续读写方式操作,现在对 X5045 中的串行 EEPROM 进行单字节的写操作和读操作,进一步熟练串行数据的读写,有兴趣者可尝试。

实验 16 串行 EEPROM 93C46 实验

一、实验目的

了解串行 EEPROM 的使用方法,掌握 16 位串行 EEPROM 数据的读写操作。

二、实验设备

STAR 系列实验仪一套、PC 机一台。

三、实验内容

1. 93C46 串行 EEPROM(C2 区)

1K bit,64×16 存储结构,根据 ORG 电位可以 8 位/16 位形式存放、读取数据(本实验中 ORG 接 VCC,以 16 位方式读写数据)。

2. 实验过程

先将数据(00H—7FH)写入 93C46,顺序写入,然后读出数据与之前写入的数据相比较,检验写入是否正确,正确:点亮 G6 区的发光管;错误:熄灭发光管。

四、实验原理图

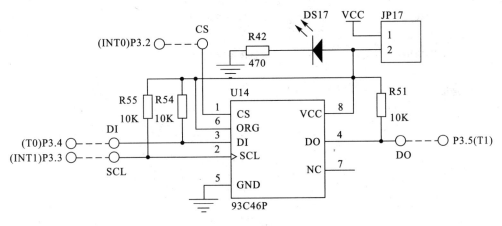

图 3-31 串行 EEPROM 接口

五、实验步骤

1. 主机连线说明:

C2 区:CS、SCL	——	A3 区:P3.2(INT0)、P3.3(INT1)
C2 区:DI、DO	——	A3 区:P3.4(T0)、P3.5(T1)
G6 区:JP65	——	A3 区:JP51(P1 口)

2. 运行程序,正确:点亮 G6 区的发光管;错误:熄灭发光管。

六、演示程序

```c
//;****************************************************************
//;93C46 串行 EEPROM 操作子程序 93C46.C
//;93C46 是 64X16 结构形式的三线串行接口串行 EEPROM,容量 1K Bit
//;****************************************************************

//控制接口
#include "reg52.h"
#include <intrins.h>

sbit CS=P3^2;              //片选,高有效
sbit SCL=P3^3;             //时钟
sbit DI=P3^4;              //数据输入
sbit DO=P3^5;              //数据输出

//指令/指令基址,指令地址要与数据所在地址相加生成操作指令
#define EWENAD0 x30        //写使能指令
#define EWDSAD0 x00        //关闭写使能指令
#define WRITEAD0 x40       //写指令基址
#define WRALAD0 x10        //整体写指令
#define READAD0 x80        //读指令基址
#define ERASEAD0 xC0       //擦除指令基址
#define ERALAD0 x20        //整体擦除指令

//初始化
void A93C46_INIT()
{
    CS=0;
    SCL=0;
    DI=0;
}

//查忙
void busy()
{
    CS=0;                  //令 CS 产生一个变化信号,DO 输出忙状态信号:0—忙
    CS=1;
    while(!DO);
}

//开始位
void Start_bit()
```

```c
{
    SCL=0;
    DI=0;
    CS=1;                          //片选,高有效
    DI=1;
    SCL=1;                         //数据在 SCL 上升沿时写入
    _nop_();
    SCL=0;
}

//一次写入 8 位数据子程序
void Write_8bits(unsigned char i)
{
    unsigned char j=0x80;
    while (j)
    {
        if (i & j)
            DI=1;
        else
            DI=0;
        SCL=1;                     //上升沿写入数据
        _nop_();
        SCL=0;
        j>>=1;
    }
}

//一次写入 16 位数据子程序,先高后低
void Write_16bits(unsigned char datal, unsigned char datah)
{
    Write_8bits(datah);            //高 8 位数据
    Write_8bits(datal);            //低 8 位数据
}

//一次读出 8 位数据子程序
unsigned char Read_8bits()
{
    unsigned char i=0;
    unsigned char j=0x80;
    
    while (j)
    {
        SCL=1;                     //高电平时读出数据
```

```c
        _nop_();
        SCL=0;
        if(DO)
            i|=j;
        j>>=1;
    }
    return i;
}

//一次读出16位数据子程序
void Read_16bits(unsigned char * pdatal, unsigned char * pdatah)
{
    * pdatah=Read_8bits();          //先读出高8位数据
    * pdatal=Read_8bits();          //后读出低8位数据
}

//写使能
void EWEN()
{
    Start_bit();                    //开始位
    Write_8bits(EWENAD);            //写使能指令:3XH
    CS=0;
}

//;关闭写使能
void EWDS()
{
    Start_bit();                    //开始位
    Write_8bits(EWDSAD);            //关闭写使能指令:0XH
    CS=0;
}

//写数据,A-操作地址
void WRITE(unsigned char address, unsigned char datal, unsigned char datah)
{
    EWEN();                         //开写使能
    Start_bit();                    //开始位
    Write_8bits(address | WRITEAD); //写操作指令基址＋数据地址——>写入操作指令
    Write_16bits(datal, datah);     //写入16位数据
    busy();                         //查忙
    CS=0;
    EWDS();                         //关写使能
}
```

```c
//整体写入
void WRAL(unsigned char address, unsigned char datal, unsigned char datah)
{
    EWEN();                              //开写使能
    Start_bit();                         //开始位
    Write_8bits(address | WRALAD);       //整体写入指令:1XH
    Write_16bits(datal, datah);          //写入 16 位数据
    busy();                              //查忙
    CS=0;
    EWDS();                              //关写使能
}

//读数据,address－操作地址
void READ(unsigned char address, unsigned char * pdatal, unsigned char * pdatah)
{
    Start_bit();                         //开始位
    Write_8bits(address | READAD);       //读操作指令基址＋数据地址－－＞读取操作指令
    Read_16bits(pdatal, pdatah);         //读出 16 位数据
    CS=0;
}

//擦除数据
void ERASE(unsigned char address)
{
    EWEN();                              //开写使能
    Start_bit();                         //开始位
    Write_8bits(address | ERASEAD);      //写入擦除数据地址
    busy();                              //查忙
    CS=0;
    EWDS();                              //关写使能
}

//整体擦除
void ERAL(unsigned char address)
{
    EWEN();                              //开写使能
    Start_bit();                         //开始位
    Write_8bits(address | ERALAD);       //整体写入指令:2XH
    busy();                              //查忙
    CS=0;
    EWDS();                              //关写使能
}
```

```c
//主程序
#define StartDATAL 02                   //起始写入数据低 8 位
#define StartDATAH 03                   //起始写入数据高 8 位

extern void A93C46_INIT();
extern void READ(unsigned char address, unsigned char * pdatal, unsigned char * pdatah);
extern void WRITE(unsigned char address, unsigned char datal, unsigned char datah);
#include "93C46.c"
#include "reg52.h"
/********************;93C46 实验子程序********************/
//;数据写入
void A93C46_Write()
{
    unsigned char datal=StartDATAL;     //起始写入数据
    unsigned char datah=StartDATAH;
    unsigned char count=64;             //共 64 次写入操作,写满整个 EEPROM
    unsigned char address=0;            //起始写入地址值
    while (count--)
    {
        WRITE(address, datal, datah);//调用数据写入子程序,93C46 是 16 位数据输入/输出形式
        address++;
        datal+=2;
        datah+=2;
    }
}
//数据检验
bit A93C46_Verify()
{
    unsigned char datal=StartDATAL;     //起始校验数据
    unsigned char datah=StartDATAH;
    unsigned char count=64;             //共 64 次操作,读整个 EEPROM
    unsigned char address=0;            //起始校验地址
    unsigned char datal1,datah1;
    while (count--)
    {
        READ(address, &datal1, &datah1);
        if ((datal!=datal1) || (datah!=datah1))
            return 0;
        address++;
        datal+=2;
        datah+=2;
    }
    return 1;
```

```
}
main()
{
    A93C46_INIT();              //93C46 初始化
    A93C46_Write();             //写入数据
    if (A93C46_Verify())        //读出数据检验
        P1=0x0;                 //显示检验结果,"正确"
    else
        P1=0xff;                //显示检验结果,"错误"
    while (1)
    {;}
}
```

第4章 拓展实验

实验1 简易电子琴实验

一、实验目的

掌握蜂鸣器的使用方法;掌握蜂鸣器不同发音的方法。

二、实验设备

STAR 系列实验仪一套、PC 机一台。

三、实验内容

1. 简易电子琴原理

①蜂鸣器输入不同频率的方波,会发出不同的声音。
②通过按键,由单片机控制产生不同频率的方波,从而发出不同的声音。

2. 实验过程

通过单片机,使 G6 区的 1~7 号键由低到高发出 1~7 的音阶。

四、实验原理图

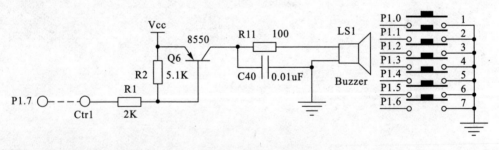

图 4-1 简易电子琴

五、实验步骤

1. 主机连线说明:

D1 区:Ctrl	——	A3 区:P1.7
G6 区:JP65	——	A3 区:JP51(P1 口)

2. 运行程序,按 G6 区的 1~7 号键,输出 7 种音阶。

3. 使用 G6 区的 1~7 号键,弹一首"生日快乐"。

六、演示程序

```
K1      BIT     P1.0            ;按键 1
K2      BIT     P1.1            ;按键 2
K3      BIT     P1.2            ;按键 3
K4      BIT     P1.3            ;按键 4
K5      BIT     P1.4            ;按键 5
K6      BIT     P1.5            ;按键 6
K7      BIT     P1.6            ;按键 7
Buzzer  BIT     P1.7            ;蜂鸣器
M1      EQU     17
M2      EQU     16
M3      EQU     15
M4      EQU     14
M5      EQU     13
M6      EQU     12
M7      EQU     11
        ORG     0000H
        LJMP    START
        ORG     0100H
START:  MOV     SP,#60H
MAIN:   MOV     P1,#0FFH        ;P1 口初始化
        LCALL   Demo            ;播放一段演示音乐
MAIN_1: JB      K1,MAIN_2       ;1 号键是否按下
        LCALL   Sound1          ;1 号键发音子程序
MAIN_2: JB      K2,MAIN_3       ;2 号键是否按下
        LCALL   Sound2          ;2 号键发音子程序
MAIN_3: JB      K3,MAIN_4       ;3 号键是否按下
        LCALL   Sound3          ;3 号键发音子程序
MAIN_4: JB      K4,MAIN_5       ;4 号键是否按下
        LCALL   Sound4          ;4 号键发音子程序
MAIN_5: JB      K5,MAIN_6       ;5 号键是否按下
        LCALL   Sound5          ;5 号键发音子程序
MAIN_6: JB      K6,MAIN_7       ;6 号键是否按下
        LCALL   Sound6          ;6 号键发音子程序
MAIN_7: JB      K7,MAIN_73      ;7 号键是否按下
        LCALL   Sound7          ;7 号键发音子程序
MAIN_73:SJMP    MAIN_1          ;返回从 1 号键开始扫描
;播放演示音乐
Demo:   MOV     DPTR,#Music
Demo_1: CLR     A
        MOVC    A,@A+DPTR
        INC     DPTR
        JZ      Demo_Ret
```

```
                LCALL       SoundMCU            ;放音子程序
                JMP         Demo_1
Demo_Ret:       RET
;乐曲:),0-结果标志
Music:
                DB          M1,M2,M3,M4,M5,M6,M7,M7,M7,M6,M5,M4,M3,M2,M1
                DB          M1,M2,M1,M2,M3,M2,M3,M4,M3,M4,M5,M4,M5,M6,M5
                DB          M6,M7,M6,M7,M7,M6,M6,M6,0
;程序控制发音
SoundMCU:       MOV         R5,#40H             ;控制放音时间
                MOV         R7,A
                ACALL       Time
                MOV         R6,A
SoundMCU_0:     PUSH        06H
SoundMCU_1:     ACALL       MusicalScale
                DJNZ        R6,SoundMCU_1
                POP         06H
                DJNZ        R5,SoundMCU_0
                RET
;计算时长
Time:           MOV         B,#0FFH
                XCH         A,B
                DIV         AB
                XCH         A,B
                CJNE        A,#8,$+3
                JC          Time_1
                INC         B
Time_1:         XCH         A,B
                RET
MusicalScale:   MOV         A,R7
MusicalScale1:  CLR         Buzzer              ;1号键发音
                LCALL       Delay               ;延时
                DJNZ        ACC,MusicalScale1
                MOV         A,R7
MusicalScale2:  SETB        Buzzer
                LCALL       Delay               ;延时
                DJNZ        ACC,MusicalScale2
                RET
;1号键发音
Sound1:         MOV         R7,#M1
Sound1_1:       ACALL       MusicalScale
                JNB         K1,Sound1_1
                RET
;2号键发音
```

```
Sound2:         MOV         R7,#M2
Sound2_1:       ACALL       MusicalScale
                JNB         K2,Sound2_1
                RET
;3号键发音
Sound3:         MOV         R7,#M3
Sound3_1:       ACALL       MusicalScale
                JNB         K3,Sound3_1
                RET
;4号键发音
Sound4:         MOV         R7,#M4
Sound4_1:       ACALL       MusicalScale
                JNB         K4,Sound4_1
                RET
;5号键发音
Sound5:         MOV         R7,#M5
Sound5_1:       ACALL       MusicalScale
                JNB         K5,Sound5_1
                RET
;6号键发音
Sound6:         MOV         R7,#M6
Sound6_1:       ACALL       MusicalScale
                JNB         K6,Sound6_1
                RET
;7号键发音
Sound7:         MOV         R7,#M7
Sound7_1:       ACALL       MusicalScale
                JNB         K7,Sound7_1
                RET
;延时
Delay:          NOP
                RET

                END
```

七、实验扩展及思考题

设计一个简易电子播放器实验程序,使用蜂鸣器,回放一段音乐。

实验 2　LED 16×16 点阵实验

一、实验目的

1. 熟悉 8155、8255 的功能,了解点阵显示的原理及控制方法。

2. 学会使用LED点阵,通过编程显示不同字符。

3. 认真预习本节实验内容,可尝试自行编写程序,做好实验准备工作,填写实验报告。

二、实验设备

STAR系列实验仪一套、PC机一台。

三、实验内容

1. 编写程序,用8255的PA、PB口控制16×16点阵的行;8255的PC口、8155的PA口控制16×16点阵的列;显示字符。

2. 按图连接线路;运行程序,观察实验结果,学会控制LED点阵显示字符。

四、实验原理图

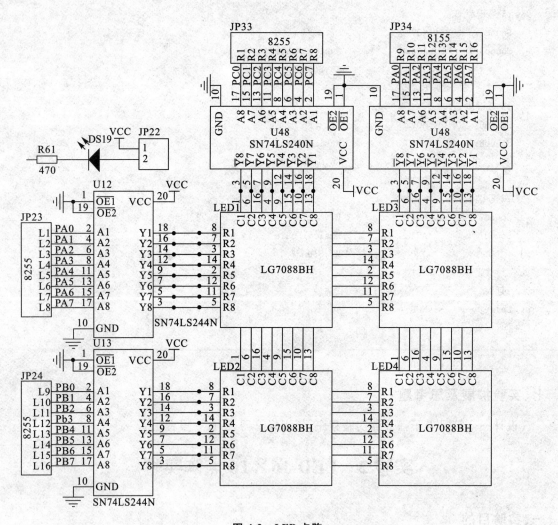

图 4-2　LED 点阵

五、实验步骤

1. 主机连线说明：

B4 区:CS(8255)、A0、A1	——	A3 区:CS1、A0、A1
B4 区:CS(8155)、IO/M		A3 区:CS2、A8
B4 区:JP56、JP53		A2 区:JP23、JP24（行输出线）
B4 区:JP52、JP76		A2 区:JP33、JP34（列输出线）

2. 运行程序，观察实验结果。运行演示程序将会看到字符"欢迎使用星研实验仪"在点阵上自下而上循环移动显示。

六、演示程序

```
            ;用 8255 的 PA、PB、PC 口和 8155 的 PA 口做 LED 16×16 点阵显示实验
A8255_PA    XDATA   0FF00H          ;8255 的 A0、A1 接总线的 A0、A1
A8255_PB    XDATA   0FF01H
A8255_PC    XDATA   0FF02H
A8255_CTL   XDATA   0FF03H
A8155_PA    XDATA   0E101H          ;8155 的 IO/M 接 P2.0
A8155_CTL   XDATA   0E100H
LINE1       XDATA   A8255_PA        ;行线 1
LINE2       XDATA   A8255_PB        ;行线 2
ROW1        XDATA   A8255_PC        ;列线 1
ROW2        XDATA   A8155_PA        ;列线 2
            ORG     0000H
            AJMP    START
            ORG     0100H
START:      MOV     SP,#60H
            CALL    INIT_IO
            CALL    TEST_LED        ;调用测试子程序,测试 LED 是否全亮
            CALL    CLEAR
;滚动显示多个字符
CHS_SHOW:   MOV     R7,#9           ;显示(R3)个字符
            MOV     DPTR,#CHAR_TAB
CHS_1:      MOV     R1,#16          ;移动 16 次
CHS_2:      ACALL   DISP_CH         ;显示一帧
            INC     DPTR
            INC     DPTR
            DJNZ    R1,CHS_2
            DJNZ    R7,CHS_1
            SJMP    CHS_SHOW
;显示一个 16×16 点阵字子程序,字形码放在 DPTR 指出的地址,显示时间为 20×R7ms
DISP_CH:    PUSH    07H
```

```
            MOV     R7,#08H
DISP_CH_1:  ACALL   DISP1
            DJNZ    R7,DISP_CH_1
            POP     07H
            RET
;显示一个 16×16 点阵字子程序,字形码放在显示缓冲区 XBUFF
DISP1:      PUSH    DPL
            PUSH    DPH
            MOV     R6,#16              ;计数器,16 列依次被扫描
            MOV     R2,#0FEH            ;上 8 行输出值
            MOV     R3,#0FFH            ;下 8 行输出值
REPEAT:     MOV     P2,#HIGH(LINE1)
            MOV     R0,#LOW(LINE1)
            MOV     A,R2
            MOVX    @R0,A               ;上 8 行输出
            MOV     R0,#LOW(LINE2)
            MOV     A,R3
            MOVX    @R0,A               ;下 8 行输出
            CLR     A
            MOVC    A,@A+DPTR
            INC     DPTR
            ACALL   ADJUST              ;调整 A,将 A 中二进制数旋转 180 度
            MOV     R0,#LOW(ROW1)
            MOVX    @R0,A               ;左边列输出
            CLR     A
            MOVC    A,@A+DPTR
            INC     DPTR
            ACALL   ADJUST              ;调整 A,将 A 中二进制数旋转 180 度
            MOV     P2,#HIGH(ROW2)
            MOV     R0,#LOW(ROW2)
            MOVX    @R0,A               ;右边列输出
            ACALL   DL10ms
            ACALL   CLEAR
            SETB    C                   ;循环移位 R2R3,行线扫描输出 0
            MOV     A,R2
            RLC     A
            MOV     R2,A
            MOV     A,R3
            RLC     A
            MOV     R3,A
            DJNZ    R6,REPEAT
            POP     DPH
            POP     DPL
            RET
```

```
;8155 和 8255 初始化
INIT_IO:    MOV     DPTR,#A8255_CTL     ;8255 控制字地址
            MOV     A,#80H              ;设置 8255 的 PA、PB、PC 口为输出口
            MOVX    @DPTR,A             ;写控制字
            MOV     DPTR,#A8155_CTL     ;8155 控制字地址
            MOV     A,#00000011B        ;设置 8155 的 PA 口为输出
            MOVX    @DPTR,A             ;写控制字
            RET
CLEAR:      MOV     A,#0FFH
            MOV     P2,#HIGH(LINE1)
            MOV     R0,#LOW(LINE1)
            MOVX    @R0,A
            MOV     R0,#LOW(LINE2)
            MOVX    @R0,A
            CLR     A
            MOV     R0,#LOW(ROW1)
            MOVX    @R0,A
            MOV     P2,#HIGH(ROW2)
            MOV     R0,#LOW(ROW2)
            MOVX    @R0,A
            RET
;调整 A 中取到的字形码的一个字节,将最高位调整为最低位,最低位调整为最高位
ADJUST:     MOV     R5,#8               ;循环移位 8 次实现
ADJUST1:    RLC     A
            XCH     A,B
            RRC     A
            XCH     A,B
            DJNZ    R5,ADJUST1
            MOV     A,B
            RET
;测试 LED 子程序,点亮 LED 并延时 1 秒
TEST_LED:   MOV     DPTR,#LINE1
            CLR     A
            MOVX    @DPTR,A
            MOV     DPTR,#LINE2
            MOVX    @DPTR,A
            MOV     DPTR,#ROW1
            MOV     A,#0FFH
            MOVX    @DPTR,A
            MOV     DPTR,#ROW2
            MOVX    @DPTR,A
            CALL    DL500ms
            CALL    DL500ms
            RET
```

```
        ;延时 10ms
DL10ms:     MOV     R4,#2
DL10ms1:    MOV     R5,#230
            DJNZ    R5,$
            DJNZ    R4,DL10ms1
            RET
DL500ms:    MOV     R5,#10
DL500ms1:   MOV     R6,#200
DL500ms2:   MOV     R7,#123
            DJNZ    R7,$
            DJNZ    R6,DL500ms2
            DJNZ    R5,DL500ms1
            RET
CHAR_TAB:
HUAN:       ;****"欢"******
            DB      00H,0C0H,00H,0C0H,0FEH,0C0H,07H,0FFH,0C7H,86H,6FH,6CH,3CH,60H,18H,60H
            DB      1CH,60H,1CH,70H,36H,0F0H,36H,0D8H,61H,9CH,0C7H,0FH,3CH,06H,00H,00H
YING:       ;****"迎"****
            DB      60H,00H,31H,0C0H,3FH,7EH,36H,66H,06H,66H,06H,66H,0F6H,66H,36H,66H
            DB      37H,0E6H,37H,7EH,36H,6CH,30H,60H,30H,60H,78H,00H,0CFH,0FFH,00H,00H
SHI:        ;****"使"***
            DB      00H,00H,06H,30H,07H,30H,0FH,0FFH,0CH,30H,1FH,0FFH,3BH,33H,7BH,33H
            DB      1BH,0FFH,1BH,33H,19H,0B0H,18H,0E0H,18H,60H,18H,0FCH,19H,8FH,1FH,03H
YONG:       ;****"用"***
            DB      0,0,1FH,0FEH,18H,0C6H,18H,0C6H,18H,0C6H,1FH,0FEH,018H,0C6H,18H,0C6H
            DB      18H,0C6H,1FH,0FEH,18H,0C6H,18H,0C6H,30H,0C6H,30H,0C6H,60H,0DEH,
0C0H,0CCH
XING:       ;****"星"***
            DB      00H,00H,1FH,0FCH,18H,0CH,1FH,0FCH,18H,0CH,1FH,0FCH,01H,80H,19H,80H
            DB      1FH,0FEH,31H,80H,31H,80H,6FH,0FCH,01H,80H,01H,80H,7FH,0FFH,00H,00H
YAN:        ;****"研"***
            DB      0,0,0FFH,0FFH,18H,0CCH,18H,0CCH,30H,0CCH,30H,0CCH,7FH,0FFH,7CH,0CCH
            DB      0FCH,0CCH,3CH,0CCH,3CH,0CCH,3DH,8CH,3DH,8CH,33H,0CH,06H,0CH,0CH,0CH
SHI:        ;****"实"***
            DB      01H,80H,00H,0C0H,3FH,0FFH,3CH,06H,67H,0CCH,06H,0C0H,0CH,0C0H,07H,0C0H
            DB      06H,0C0H,7FH,0FFH,00H,0C0H,01H,0E0H,03H,30H,06H,18H,1CH,1CH,70H,18H
YANO:       ;****"验"***
            DB      00H,00H,0FCH,60H,0CH,60H,6CH,0F0H,6CH,0D8H,6DH,8FH,6FH,0F8H,7EH,00H
            DB      06H,0C6H,07H,66H,3FH,0ECH,0E7H,0ECH,06H,18H,1FH,0FFH,0CH,00H,00H,00H
YI:         ;****"仪"***
            DB      0CH,0C0H,0CH,60H,18H,7CH,1BH,6CH,33H,0CH,73H,18H,0F1H,98H,31H,98H
            DB      30H,0F0H,30H,0F0H,30H,60H,30H,0F0H,31H,98H,33H,0FH,3EH,06H,30H,00H
NONE:       ;送暗码(不显示)
            DB      00H,00H,00H,00H,00H,00H,00H,00H,00H,00H,00H,00H,00H,00H,00H,00H
```

```
DB    00H,00H,00H,00H,00H,00H,00H,00H,00H,00H,00H,00H,00H,00H,00H,00H
END
```

七、实验扩展及思考

1. 如果并行扩展口线不多,能不能用串并转换方式替代 8255 和 8155？若可以,具体如何连线？若不行,还有其他方法吗？

2. 修改程序,使显示的字符从左至右动态循环显示。

实验 3　I²C 总线串行 EEPROM 24C02A 实验

一、实验目的

了解 I²C 总线读写方式；掌握 I²C 总线的读写操作和对 24C02A 进行数据读写。

二、实验设备

STAR 系列实验仪一套、PC 机一台。

三、实验内容

1. 24C02A

①24C02A 是 I²C 总线(二线串行接口)的串行 EEPROM,容量 4K bits。
②模式分为字节写和页写(8 字节),可以单字节读取或连续读出数据。

2. 实验过程

①写满 24C02A 内部整个 4K bits 串行 EEPROM,然后检验写入数据是否正确并显示结果,正确:点亮 8 个红色发光管(G6 区),错误:熄灭 8 个红色发光管。
②起始写入地址为 00H,起始写入数据为 00H,之后地址与数据都以＋1 递增,直到写满整个 EEPROM。

四、实验原理图

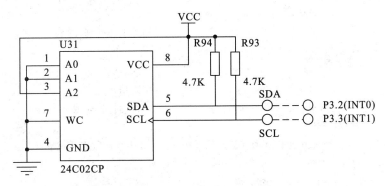

图 4-3　EEPROM

五、实验步骤

1. 主机连线说明：

E4 区：SDA	——	A3 区：P3.2(INT0)
E4 区：SCL	——	A3 区：P3.3(INT1)
G6 区：JP65	——	A3 区：JP51(P1)

2. 运行程序：向 24C02A 写入数据，然后读出数据检验，检验正确，8 个发光管（G6 区）全亮；检验错误，8 个发光管（G6 区）全灭。

六、演示程序

```
;1.24C02A 子程序(24C02A.ASM)
SDA             BIT     P3.2            ;数据传输口
SCL             BIT     P3.3            ;时钟
;24C02 的片选地址:100H
A24C02_WRITE    EQU     0A8H            ;写指令
A24C02_READ     EQU     0A9H            ;读指令
;初始化
A24C02_INIT:    SETB    SCL             ;将 SCL,SDA 置位,释放 IIC 总线
                SETB    SDA
                RET
;写操作,分字节写和页写模式
;字节写,一次写入一个字节数据,A—写入地址,B—数据
Write_Byte:     PUSH    ACC             ;A 中地址压栈
                MOV     A,#A24C02_WRITE ;写操作指令
                LCALL   Start           ;开始信号
                LCALL   Write_8bits     ;写入 8 位数据
                LCALL   Acknowledge     ;查询接收端应答信号
                POP     ACC             ;写入 A 中地址
                LCALL   Write_8bits
                LCALL   Acknowledge
                MOV     A,B             ;写入 B 中数据
                LCALL   Write_8bits
                LCALL   Acknowledge
                LCALL   Stop            ;结束信号
                LCALL   AckPolling      ;等待写操作完成
                RET
;页写,一次写入 8 个字节数据,A 中存放起始写入地址,R0 中存放数据首地址
Write_Page:     PUSH    07H
                MOV     R7,#8
                PUSH    ACC             ;A 中地址压栈
                MOV     A,#A24C02_WRITE ;写操作指令
```

```
                LCALL       Start               ;开始信号
                LCALL       Write_8bits         ;写入8位数据
                LCALL       Acknowledge         ;查询接收端应答信号
                POP         ACC                 ;写入A中地址
                LCALL       Write_8bits
                LCALL       Acknowledge
                PUSH        ACC
Write_Page_1:   MOV         A,@R0               ;缓存区数据逐一写入
                LCALL       Write_8bits
                LCALL       Acknowledge
                INC         R0
                DJNZ        R7,Write_Page_1     ;写8次
                LCALL       Stop                ;结束信号
                CLR         A
                LCALL       AckPolling          ;等待写操作完成
                POP         ACC
                POP         07H
                RET
;等待写操作完成
AckPolling:     MOV         A,#A24C02_Write     ;写操作指令
                LCALL       Start               ;开始信号
                LCALL       Write_8bits
                SETB        SDA                 ;从机应答
                SETB        SCL                 ;应答占用一个脉冲
                LCALL       Delay_Time
                JB          SDA,AckPolling      ;接收端应答标志:将SDA置低
                CLR         SCL
                LCALL       Stop                ;停止信号
                RET
;读操作,分为字节读和连续读取操作
;字节读,一次读取一个字节,A—读取地址
Read_Byte:      PUSH        ACC                 ;A中地址压栈
                LCALL       Start               ;开始信号
                MOV         A,#A24C02_Write     ;写操作指令
                LCALL       Write_8bits
                LCALL       Acknowledge
                POP         ACC                 ;写入A中地址
                LCALL       Write_8bits
                LCALL       Acknowledge
;立即读,读取当前内部地址的数据,一个字节
Read_Current:   LCALL       Start               ;开始信号,下面读取数据
                MOV         A,#A24C02_Read      ;读操作指令
                LCALL       Write_8bits
```

```
                    LCALL       Acknowledge
                    LCALL       Read_8bits              ;读取数据,放在 A 中
                    LCALL       Stop                    ;停止信号
                    RET
;连续读取 n 个数据,A—读取首地址,B—存放读取数据个数
;R0—缓冲区
Read_Sequence:      PUSH        07H
                    PUSH        ACC
                    DEC         B
                    MOV         R7,B                    ;B 中存放读取数据个数
                    LCALL       Start                   ;开始信号
                    MOV         A,#A24C02_Write         ;写操作指令
                    LCALL       Write_8bits
                    LCALL       Acknowledge
                    POP         ACC
                    LCALL       Write_8bits
                    LCALL       Acknowledge
                    LCALL       Start                   ;开始信号,下面读取数据
                    MOV         A,#A24C02_Read          ;读操作指令
                    LCALL       Write_8bits
                    LCALL       Acknowledge
Read_Sequence_1:    LCALL       Read_8bits
                    LCALL       MasterACK
                    MOV         @R0,A                   ;数据存到 R0 指向的 RAM 中
                    INC         R0
                    DJNZ        R7,Read_Sequence_1
                    LCALL       Read_8bits              ;最后一次读无应答
                    MOV         @R0,A
                    LCALL       Stop                    ;停止信号
                    POP         07H
Read_Sequence_2:    RET
;写入 8 位数据
Write_8bits:        PUSH        07H
                    MOV         R7,#8
Write_8bits_1:      RLC         A
                    CLR         SCL                     ;数据在 SCL 为低时 SDA 上的数据可以改变,此时送上欲写数据
                    LCALL       Delay_Time              ;延时
                    MOV         SDA,C
                    SETB        SCL
                    LCALL       Delay_Time
                    DJNZ        R7,Write_8bits_1
                    CLR         SCL
                    POP         07H
```

```
                    RET
;读取 8 位数据
Read_8bits:         PUSH        07H
                    MOV         R7,#8
Read_8bits_1:       CLR         SCL
                    LCALL       Delay_Time
                    SETB        SCL                     ;高电平读出数据
                    MOV         C,SDA
                    RLC         A
                    DJNZ        R7,Read_8bits_1
                    CLR         SCL
                    POP         07H
                    RET
;开始信号
Start:              SETB        SDA;I²C 总线操作开始信号:SCL 为高时,SDA 由高→低
                    SETB        SCL
                    LCALL       Delay_Time
                    CLR         SDA
                    LCALL       Delay_Time
                    RET
;结束信号
Stop:               CLR         SDA ;I²C 总线操作结束信号:SCL 为高时,SDA 由低→高
                    SETB        SCL
                    LCALL       Delay_Time
                    SETB        SDA                     ;结束操作,将 SCL、SDA 置高,释放总线
                    LCALL       Delay_Time
                    RET
;应答查询
;从机应答
Acknowledge:        SETB        SDA                     ;从机应答
                    SETB        SCL                     ;应答占用一个脉冲
                    LCALL       Delay_Time
                    JB          SDA,$                   ;接收端应答标志:将 SDA 置低
                    CLR         SCL
                    RET
;主机应答
MasterACK:          CLR         SDA                     ;数据线 SDA 清 0 应答
                    SETB        SCL                     ;产生一个脉冲令从机接收到应答
                    LCALL       Delay_Time
                    CLR         SCL
                    SETB        SDA                     ;必须置高数据
                    RET
;延时
```

```
Delay_Time:         RET

                    END
;2.主程序(MAIN.ASM)
;写入数据,256字节串行EEPROM顺序写入00H—0FFH
A24C02_Write:       MOV     R7,#32              ;32次页写,每次页写写入8个字节,共256个字节
                    MOV     R3,#00H             ;写入首地址
                    MOV     R2,#VERIFYDATA      ;起始写入数据
A24C02_Write_1:     MOV     R0,#buffer          ;写入数据先放在buffer(30H开始的内部RAM)
A24C02_Write_2:     MOV     @R0,02H
                    INC     R0
                    INC     R2
                    CJNE    R0,#buffer+8,A24C02_Write_2 ;一页写入8个字节
                    MOV     R0,#buffer
                    MOV     A,R3
                    LCALL   Write_Page
                    MOV     A,R3
                    ADD     A,#8
                    MOV     R3,A
                    DJNZ    R7,A24C02_Write_1
                    RET
;检验数据,读出数据与写入数据一一对应相比较,检验写入是否正确
                    MOV     R7,#0FFH            ;读取整个EEPROM内的数据,256个字节
                    MOV     R1,#buffer
                    MOV     R2,#VERIFYDATA      ;数据检验
                    MOV     B,#00H              ;检验EEPROM起始数据地址
A24C02_Verify_1:    MOV     A,B
                    LCALL   Read_Byte           ;读取数据
                    XCH     A,R1
                    CJNE    A,#buffer+30H,$+3   ;写入片内RAM,超过30H个字节,停止写入
                    XCH     A,R1
                    JNC     A24C02_Verify_3
                    MOV     @R1,A               ;读出的数据顺序写入片内RAM,便于检查
                    INC     R1
A24C02_Verify_3:    CJNE    A,02H,A24C02_Verify_2
                    INC     R2
                    INC     B
                    DJNZ    R7,A24C02_Verify_1
                    CLR     F0                  ;F0为数据检验结果标志,0—正确
                    RET
A24C02_Verify_2:    SETB    F0                  ;1—检验错误
                    RET
```

七、实验扩展及思考题

学会使用 24C02A 的其余指令,如字节写入、连续读取等,进一步熟悉 I^2C 总线操作。

实验 4 电子钟(PCF8563(I^2C 总线)、128×64 液晶)

一、实验目的

进一步熟悉 I^2C 总线;掌握时钟芯片的使用;掌握使用液晶显示器显示时间。

二、实验设备

STAR 系列实验仪一套、PC 机一台。

三、实验内容

1. PCF8563

① 实时时钟芯片,可计时时间 1900—2099 年,不具有周调整功能。
② 数据传输采用 I^2C 总线,固定片选地址;闹铃中断功能,可编程频率输出。

2. 实验过程

读写 PCF8563 中的时间数据;在 12864J 液晶显示器上显示时间、星期、日期。

四、实验原理图

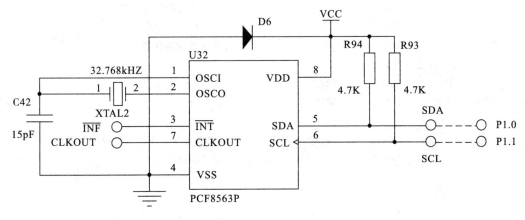

图 4-4 PCF8563 接口

五、实验步骤

1. 主机连线说明:

E4 区:SDA	——	A3 区:P1.0
E4 区:SCL	——	A3 区:P1.1
A1 区:CS,RW,RS,CS1/2	——	A3 区:CS1,A0,A1,A2

2. 初始化 PCF8563,设置初始化时间(2005-07-01 Fri 12:30:00),读取时间数据。

3. 调整读取的时间数据,转换为可以在图形液晶显示器上显示的数据,显示时间(年、月、日、星期、小时、分、秒)。

六、演示程序

1. PCF8563 子程序(PCF8563.ASM)

SDA	BIT	P1.0	;数据
SCL	BIT	P1.1	;时钟

;内部寄存器地址

CS1	EQU	00H	;控制/状态寄存器 1
CS2	EQU	01H	;控制/状态寄存器 2
Second	EQU	02H	;秒寄存器
Minute	EQU	03H	;分寄存器
Hour	EQU	04H	;时寄存器
Day	EQU	05H	;日寄存器
Weekday	EQU	06H	;周寄存器
Mouth	EQU	07H	;月寄存器
Year	EQU	08H	;年寄存器
MinuteA	EQU	09H	;分闹铃寄存器
HourA	EQU	0AH	;时闹铃寄存器
DayA	EQU	0BH	;日闹铃寄存器
WeekdayA	EQU	0CH	;周闹铃寄存器
CO	EQU	0DH	;时钟输出控制寄存器
TimerCtrl	EQU	0EH	;定时控制寄存器
Timer	EQU	0FH	;定时设置寄存器

;PCF8563 的片选地址:001H

PCF8563_WRITE	EQU	0A2H	;写指令
PCF8563_READ	EQU	0A3H	;读指令

;初始化

PCF8563_INIT:	LCALL	Delay8ms	;I^{2}C 总线从通电到开始操作要求 8ms 延时
	LCALL	Register_INIT	;寄存器初始化
INIT_1:	RET		

;寄存器初始化

Register_INIT:	LCALL	CS1_Set	;设置控制/状态寄存器 1
	LCALL	CS2_Set	;设置控制/状态寄存器 2
	LCALL	DayA_Set	;设置天闹铃寄存器
	LCALL	WeekdayA_Set	;设置周闹铃寄存器
	LCALL	CO_Set	;设置时钟输出寄存器
	RET		
CS1_Set:	MOV	A,#CS1	;A—寄存器地址,B—寄存器设置值
	MOV	B,#00H	;全设为正常模式
	LCALL	Write	;数据写入子程序

```
                RET
CS2_Set:        MOV     A,#CS2
                MOV     B,#02H          ;开闹铃中断,关定时器中断
                LCALL   Write
                RET
DayA_Set:       MOV     A,#DayA
                MOV     B,#00H          ;关日闹铃
                LCALL   Write
                RET
WeekdayA_Set:   MOV     A,#WeekdayA
                MOV     B,#00H          ;关周闹铃
                LCALL   Write
                RET
CO_Set:         MOV     A,#CO
                MOV     B,#00H          ;关时钟输出
                LCALL   Write
                RET
;读秒,返回 A—秒数据
Sec_Read:       MOV     A,#Second
                LCALL   Read
                ANL     A,#7FH
                RET
;读取时间,TIME—时间数据缓冲区
Time_Read:      MOV     A,#Second       ;起始读取地址,从秒开始
                MOV     R7,#7           ;连续读取数据个数:秒,分,时,日,周,月,年
                MOV     R0,#TIME        ;时间数据缓冲区
                LCALL   Read_Sequence   ;连续读取子程序
                LCALL   Time_Adjust
                RET
;时间写入,将秒,分,时,日,周,月,年写入;TIME—欲写入时间数据缓冲区
Time_Write:     MOV     R0,#TIME        ;时间数据缓冲区首地址
                MOV     A,#Second       ;秒数据地址为起始写入地址
Time_Write_1:   PUSH    ACC
                MOV     B,@R0           ;B—时间数据
                LCALL   Write
                POP     ACC
                INC     A
                INC     R0
                CJNE    R0,#TIME+7,Time_Write_1
                RET
;写入1个字节数据,A—寄存器地址,B—数据
Write:          PUSH    ACC             ;寄存器地址压栈
                LCALL   Start           ;开始信号
```

```
                    MOV         A,#PCF8563_WRITE    ;写操作指令
                    LCALL       Write_8bits         ;写入8位数据
                    LCALL       Acknowledge         ;查询接收端应答信号
                    POP         ACC                 ;写入寄存器地址
                    LCALL       Write_8bits
                    LCALL       Acknowledge
                    MOV         A,B                 ;写入设置值
                    LCALL       Write_8bits
                    LCALL       Acknowledge
                    LCALL       Stop                ;停止信号
                    RET
;读取数据,分为两种模式:字节读取和连续读取
;字节读取,一次读取一个字节的数据,A—读取地址及存放读出的数据
Read:               PUSH        ACC                 ;寄存器地址压栈
                    LCALL       Start               ;开始信号
                    MOV         A,#PCF8563_WRITE    ;写操作指令
                    LCALL       Write_8bits         ;写入8位数据
                    LCALL       Acknowledge         ;查询接收端应答信号
                    POP         ACC                 ;写入寄存器地址
                    LCALL       Write_8bits
                    LCALL       Acknowledge
                    LCALL       Start
                    MOV         A,#PCF8563_Read     ;读操作指令
                    LCALL       Write_8bits
                    LCALL       Acknowledge
                    LCALL       Read_8bits          ;读取数据
                    LCALL       Stop                ;停止信号
                    RET
;连续读取n个数据,n≤16;A—寄存器首地址,R7—读取数据个数,R0—读取数据存放首地址
Read_Sequence:      PUSH        ACC                 ;寄存器地址压栈
                    LCALL       Start               ;开始信号
                    MOV         A,#PCF8563_WRITE    ;写操作指令
                    LCALL       Write_8bits         ;写入8位数据
                    LCALL       Acknowledge         ;查询接收端应答信号
                    POP         ACC                 ;写入寄存器地址
                    LCALL       Write_8bits
                    LCALL       Acknowledge
                    LCALL       Start
                    MOV         A,#PCF8563_Read     ;读操作指令
                    LCALL       Write_8bits
                    LCALL       Acknowledge
                    SJMP        Read_Sequence_1
Read_Sequence_2:    LCALL       MasterACK
```

```
Read_Sequence_1:  LCALL      Read_8bits           ;读取数据
                  MOV        @R0,A
                  INC        R0
                  DJNZ       R7,Read_Sequence_2
                  LCALL      Stop
                  RET
Delay_tHD_DAT     MACRO
                  NOP
                  ENDM
Delay_tSU_DAT     MACRO
                  NOP
                  ENDM
Delay_tHIGH       MACRO
                  NOP
                  NOP
                  ENDM
Delay_tLOW        MACRO
                  NOP
                  ENDM
Delay_tHD_STA     MACRO
                  NOP
                  ENDM
Delay_tSU_STO     MACRO
                  NOP
                  ENDM
Delay_tBUF        MACRO
                  NOP
                  NOP
                  ENDM
;写入8位数据
Write_8bits:      PUSH       07H
                  MOV        R7,#8
Write_8bits_1:    CLR        SCL
                  Delay_tHD_DAT
                  RLC        A
                  MOV        SDA,C                ;在 SCL 为低时,将数据送上 SDA
                  Delay_tSU_DAT
                  SETB       SCL
                  Delay_tHIGH
                  DJNZ       R7,Write_8bits_1
                  CLR        SCL
                  POP        07H
                  RET
```

```
;读取8位数据
Read_8bits:     PUSH        07H
                MOV         R7,#8
Read_8bits_1:   CLR         SCL
                Delay_tLOW
                SETB        SCL             ;SCL 高电平时,读取 SDA 数据
                Delay_tHIGH
                MOV         C,SDA
                RLC         A
                DJNZ        R7,Read_8bits_1
                CLR         SCL
                POP         07H
                RET
;开始信号
Start:          SETB        SDA             ;I²C 总线操作开始信号:SCL 为高时,SDA 由高→低
                SETB        SCL
                Delay_tBUF  CLRSDA
                Delay_tHD_STA
                CLR         SCL
                RET
;结束信号
Stop:           Delay_tHD_DAT
                CLR         SDA             ;I²C 总线操作结束信号:SCL 为高时,SDA 由低→高
                SETB        SCL
                Delay_tSU_STO
                SETB        SDA             ;操作结束后,确保 I²C 总线处于释放状态
                RET
;从机应答查询
Acknowledge:    Delay_tLOW
                SETB        SDA             ;查询接收端应答信号,要先释放总线
                SETB        SCL
                Delay_tHIGH
                JB          SDA,$           ;接收端应答标志:将 SDA 置低
                CLR         SCL
                RET
;主机应答
MasterACK:      Delay_tHD_DAT
                CLR         SDA             ;数据线 SDA 置 0 应答
                Delay_tSU_DAT
                SETB        SCL
                Delay_tHIGH
                CLR         SCL
                SETB        SDA             ;置高数据
                RET
```

2. 主程序(MAIN.ASM)说明

①初始化 PCF8563,写入初始化时间。因为没有给 PCF8563 配置电池,所以对 PCF8563 初始化后,给它赋时间初值,才能正确显示。

②读取时间,调整为非压缩 BCD 码,显示在液晶显示器 12864J 上。液晶显示程序请参阅基础实验 18。

七、实验扩展及思考题

使用 PCF8563,输出一个 20kHz 的方波、提供闹钟功能,有兴趣者可自行尝试。

实验 5 电子钟(CLOCK)

一、实验目的

进一步熟悉 I^2C 总线;掌握 I^2C 总线多芯片操作;掌握时钟芯片的使用;掌握使用液晶显示器;制作一个简易时钟。

二、实验设备

STAR 系列实验仪一套、PC 机一台。

三、实验内容

1. PCF8563

①实时时钟芯片,可计时时间 1900—2099 年,不具有周调整功能。

②数据传输采用 I^2C 总线,固定片选地址;闹铃中断功能,可编程频率输出。

2. 实验过程

读写 PCF8563 中的时间数据;在 12864J 液晶显示器上显示时间;用 G5 区的 0、1、2 按键设置时间:0 键——时间+1;1 键——时间-1;2 键——更改时间设置光标。

四、实验原理图

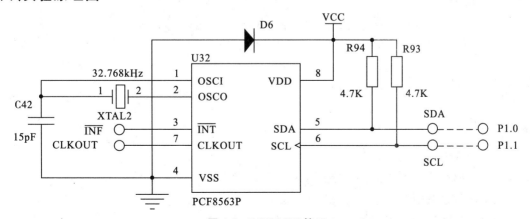

图 4-5 PCF8563P 接口

五、实验步骤

1. 主机连线说明：

E4 区:SDA	——	A3 区:P1.0
E4 区:SCL	——	A3 区:P1.1
E4 区:KEY	——	A3 区:P3.2(INT0)
A1 区:CS、RW、RS、CS1/2	——	A3 区:CS1、A0、A1、A2

2. 读取时间数据，经过转换调整时间数据，显示在液晶显示器上。

3. 通过按键设置时间为当前北京时间，并显示在液晶显示器(年、月、日和小时、分、秒以及星期)。

六、演示程序

1. ZLG7290、12864 液晶、PCF8563 程序请参阅前边实验。

2. 主程序(MAIN.ASM)略。主程序说明：
①PCF8563 初始化。
②读取时间，调整时间数据，显示在液晶显示器上。
③通过 1、2、3 号键(G6 区)设置时间，设置时间时，当前设置的时间会闪烁。显示、设置时间程序较长，但并不复杂，请参阅 CLOCK\MAIN.ASM。

七、实验扩展及思考题

1. 实验名称：闹钟实验。
2. 实验内容：利用 PCF8563 的闹钟功能，在已做好的上面的实验基础上，增加闹钟功能。

实验 6 数字式温度计实验(18B20、ZLG7290)

一、实验目的

掌握一线串行接口的读写操作；掌握数字温度计 DS18B20 的使用。

二、实验设备

STAR 系列实验仪一套、PC 机一台。

三、实验内容

1. DS18B20

①一线串行接口数字式温度计。
②温度测量范围-55~125℃，-10~85℃内误差±0.5℃。
③9~12 位转换精度，转换时间 100~750ms，通常为 500ms。

2. 实验过程

应用 DS18B20 制作一个数字温度计,通过 DS18B20 测量温度,ZLG7290 控制 LED(G5 区)动态显示温度。

四、实验原理图

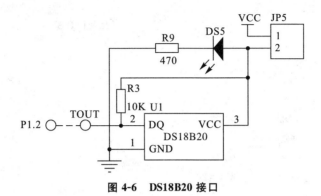

图 4-6 DS18B20 接口

五、实验步骤

1. 主机连线说明:

G1 区:TOUT	——	A3 区:P1.2
E4 区:SDA	——	A3 区:P1.0
E4 区:SCL	——	A3 区:P1.1
E4 区:A、B、C、D	——	G5 区:A、B、C、D

2. 使用 DS18B20 测量温度,将读出的十六进制温度值转换为十进制数。

3. 通过 LED(G5 区)动态显示温度,温度数据通过 DS18B20 获取。可用手指贴住 DS18B20(G1 区),温度显示会随之上升。

六、演示程序

1. 18B20 子程序

```
DQ              BIT     P1.2            ;数据输入/输出端
;DS18B20复位初始化子程序
INIT_18B20:     SETB    DQ
                NOP
                CLR     DQ              ;主机发出501us的复位低脉冲
                MOV     R7,#250
                DJNZ    R7,$
                SETB    DQ              ;拉高数据线
                MOV     R7,#30
TSR:            JNB     DQ,TSR1         ;等待DS18B20回应
                DJNZ    R7,TSR
                SETB    C               ;置位标志位,表示DS18B20不存在
```

```
                    RET
TSR1:               CLR     C                   ;复位标志位,表示 DS18B20 存在
                    MOV     R7,#250
                    DJNZ    R7,$                ;时序要求延时一段时间
                    RET
;写操作
WRITE_18B20:        MOV     R7,#8               ;一共 8 位数据
                    CLR     C
WRI:                NOP
                    CLR     DQ
                    MOV     R6,#3
                    DJNZ    R6,$
                    RRC     A
                    MOV     DQ,C
                    MOV     R6,#26
                    DJNZ    R6,$
                    SETB    DQ
                    DJNZ    R7,WRI
                    RET
;读操作
READ_18B20:         MOV     R7,#8               ;数据一共有 8 位
READ1:              CLR     DQ
                    NOP
                    SETB    DQ
                    MOV     R6,#3
                    DJNZ    R6,$
                    MOV     C,DQ
                    MOV     R6,#26
                    DJNZ    R6,$
                    RRC     A
                    DJNZ    R7,READ1
                    RET
;判断 DS18B20 是否存在,启动 DS18B20             ;CY 为判断标志
START_Temperature:  SETB    DQ
                    ACALL   INIT_18B20          ;先复位 DS18B20
                    JC      GET_T
                    MOV     A,#0CCH             ;跳过 ROM 匹配
                    LCALL   WRITE_18B20
                    MOV     A,#44H              ;发出温度转换命令
                    LCALL   WRITE_18B20
                    CLR     C
GET_T:              RET
;读出转换后的温度值,保存于:A—高 8 位数据,B—低 8 位数据
```

RD_Temperature:	LCALL	INIT_18B20	;准备读温度前先复位
	MOV	A,#0CCH	;跳过ROM匹配
	LCALL	WRITE_18B20	
	MOV	A,#0BEH	;发出读温度命令
	LCALL	WRITE_18B20	
	CALL	READ_18B20	;读出温度
	MOV	B,A	;存放到A,B中
	CALL	READ_18B20	
	RET		

2. 主程序(MAIN.ASM)

;主程序说明：
;向DS18B20发出温度转换信号,延时等待,读出温度;将温度值由十六进制转换成十进
;制数,使用ZLG7290控制LED显示温度

TEMPER_L:	DS	1	;用于保存读出温度的低8位
TEMPER_H:	DS	1	;用于保存读出温度的高8位
buffer:	DS	8	;温度临时存放区
	...		
MAIN:	LCALL	START_Temperature	;向DS18B20发送读温度指令
	LCALL	DelayTime	
	LCALL	RD_Temperature	;读出温度值,并转换为BCD码
	MOV	TEMPER_L,B	;温度个位,小数位数据
	MOV	TEMPER_H,A	;温度十位数据
	LCALL	DIS_BCD	;提取温度数据,转换为非压缩型BCD码,并显示
	SJMP	MAIN	

;温度转换/显示

DIS_BCD:	MOV	R0,#buffer+3	;设置显示内容存放区首地址
	MOV	@R0,#10H	;正数
	MOV	A,TEMPER_H	
	JNB	ACC.3,DIS_BCD1	;判断温度是否为负
	MOV	@R0,#11H	;负数
	CPL	A	
	XCH	A,TEMPER_L	
	CPL	A	
	ADD	A,#1	
	XCH	A,TEMPER_L	
	ADDC	A,#0	
DIS_BCD1:	ANL	A,#0FH	;将温度整数位转换为ASCII
	MOV	B,A	
	MOV	A,TEMPER_L	
	ANL	A,#0F0H	
	ORL	A,B	;将温度的个位与十位BCD合在一起
	SWAP	A	
	MOV	B,#10	

```
            DIV     AB
            JNZ     DIS_BCD2            ;判断温度的十位是否为0,并进行相应处理
            MOV     A,#10H              ;十位为0
            XCH     A,@R0
            DEC     R0
            MOV     @R0,A
            SJMP    DIS_BCD3
DIS_BCD2:   DEC     R0
            MOV     @R0,A
DIS_BCD3:   DEC     R0
            MOV     @R0,B
            DEC     R0
            MOV     A,TEMPER_L          ;转换小数部分
            ANL     A,#0FH
            MOV     B,A
            CLR     A
            JNB     B.0,DIS_BCD4
            MOV     A,#6
DIS_BCD4:   JNB     B.1,DIS_BCD5
            ADD     A,#12H
            DA      A
DIS_BCD5:   JNB     B.2,DIS_BCD6
            ADD     A,#25H
            DA      A
DIS_BCD6:   JNB     B.3,DIS_BCD7
            ADD     A,#50H
            DA      A
DIS_BCD7:   SWAP    A
            ANL     A,#0FH
            MOV     @R0,A
            MOV     R0,#buffer+4        ;显示数据首地址
            MOV     @R0,#10H
            INC     R0
            MOV     @R0,#10H
            INC     R0
            MOV     @R0,#10H
            INC     R0
            MOV     @R0,#10H
            MOV     R0,#buffer
            LCALL   Display8
            RET
```

七、实验扩展及思考题

读取 DS18B20 内部 64 位识别码,了解多个 DS18B20 协同工作原理。

实验 7　步进电机实验

一、实验目的

1. 了解步进电机的基本原理,掌握步进电机的转动编程方法。
2. 了解影响电机转速的因素。

二、实验设备

STAR 系列实验仪一套、PC 机一台。

三、实验内容

编写程序:使用 G5 区的键盘控制步进电机的正反转、调节转速,连续转动或转动指定步数;将相应的数据显示在 G5 区的数码管上。

四、控制原理

步进电机的驱动原理是通过它每相线圈的电流的顺序切换来使电机作步进式旋转,驱动电路由脉冲来控制,所以调节脉冲的频率便可改变步进电机的转速,微控制器最适合控制步进电机。另外,由于电机的转动惯量的存在,其转动速度还受驱动功率的影响,当脉冲的频率大于某一值(本实验为 $f_.>100Hz$)时,电机便不再转动。

实验电机共有 4 个相位(A,B,C,D),按转动步骤可分单 4 拍(A—>B—>C—>D—>A)、双 4 拍(AB—>BC—>CD—>DA—>AB)和单双 8 拍(A—>AB—>B—>BC—>C—>CD—>D—>DA—>A)。

五、实验原理图

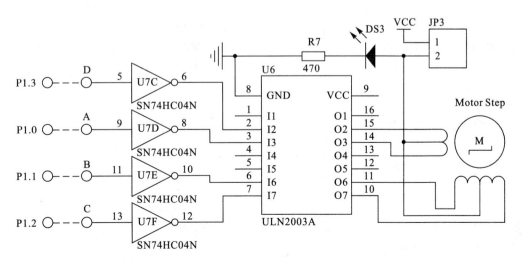

图 4-7　步进电机接口

六、实验步骤

1. 主机连线说明：

E1 区:A、B、C、D	——	A3 区:P1.0、P1.1、P1.2、P1.3
E5 区:CLK	——	B2 区:2M
E5 区:CS、A0		A3 区:CS5、A0
E5 区:A、B、C、D	——	G5 区:A、B、C、D

2. 调试程序，查看运行结果是否正确。

七、演示程序

```
                    NAME        MAIN                        ;定义模块名
        EXTRN       CODE(INIT8279,SCAN_KEY,Display8)        ;8279.ASM 中定义的子程序
        MAIN_CODE   SEGMENT     CODE
        MAIN_BIT    SEGMENT     BIT
        MAIN_DATA   SEGMENT     DATA
        STACK       SEGMENT     IDATA

                    RSEG        MAIN_DATA
        StepControl:    DS      1           ;下一次送给步进电机的值
        buffer:         DS      8           ;显示缓冲区,8 个字节
        SpeedNo:        DS      1           ;选择哪一级速度
        StepDelay:      DS      1           ;转动一步后,延时常数
        ;如果选择的速度快于启动速度,延时由长到短,最终使用对应的延时常数
        StartStepDelay: DS      1
        StartStepDelay1:DS      1           ;StartStepDelay

                    RSEG        MAIN_BIT
        bFirst:         DBIT    1           ;有没有转动过步进电机
        bClockwise:     DBIT    1           ;=1 顺时针方向,=0 逆时针方向转动
        bNeedDisplay:   DBIT    1           ;已转动一步,需要显示新步数
                    RSEG        STACK
                        DS      20H         ;32Bytes Stack
                    CSEG        AT  0000H   ;定位 0
                        LJMP    STAR
                    CSEG        AT  000BH
                        LJMP    TIMER0
                    RSEG        MAIN_CODE
        STAR:           MOV     SP,#STACK-1
                        ACALL   INIT8279
                        SETB    bFirst
                        SETB    bClockwise
```

```
                MOV     StepControl,#33H   ;下一次送给步进电机的值
                MOV     SpeedNo,#5
                MOV     TMOD,#02H
                MOV     TH0,#55
                MOV     TL0,#55            ;200us 延时
                MOV     IE,#82H
                MOV     Buffer+7,#0
                MOV     buffer+6,#0
                MOV     buffer+5,#0
                MOV     buffer+4,#0
                MOV     buffer+3,#10H
                MOV     buffer+2,SpeedNo
                MOV     buffer+1,#10H
                MOV     buffer,#0
STAR2:          MOV     R0,#buffer
                ACALL   Display8
STAR3:          ACALL   Scan_Key
                JC      STAR5
                JNB     bNeedDisplay,STAR3
                CLR     bNeedDisplay
                ACALL   Step_SUB_1
                SJMP    STAR2
STAR5:          CLR     TR0                ;终止步进电机转动
                CJNE    A,#10,$+3
                JNC     STAR1
                MOV     buffer+4,buffer+5
                MOV     buffer+5,buffer+6
                MOV     buffer+6,buffer+7
                MOV     Buffer+7,A
                SJMP    STAR2
STAR1:          CJNE    A,#14,$+3
                JNC     STAR3
                MOV     DPTR,#DriverTab
                CLR     C
                SUBB    A,#10
                RL      A
                JMP     @A+DPTR
DriverTab:      SJMP    Direction          ;转动方向
                SJMP    Speed_up           ;提高转速
                SJMP    Speed_Down         ;降低转速
                SJMP    Exec               ;步进电机根据方向、转速、步数开始转动
Direction:      CPL     bClockwise
                JB      bClockwise,Clockwise
```

```
                MOV      buffer,#1
AntiClockwise:  JNB      bFirst,AntiClockwise1
                MOV      StepControl,#91H
                SJMP     Direction1
AntiClockwise1: MOV      A,StepControl
                RR       A
                RR       A
                MOV      StepControl,A
                SJMP     Direction1
Clockwise:      MOV      buffer,#0
                JNB      bFirst,Clockwise1
                MOV      StepControl,#33H
                SJMP     Direction1
Clockwise1:     MOV      A,StepControl
                RL       A
                RL       A
                MOV      StepControl,A
Direction1:     SJMP     STAR2
Speed_up:       MOV      A,SpeedNo
                CJNE     A,#11,Speed_up1
                SJMP     Speed_up2
Speed_up1:      INC      SpeedNo
                MOV      buffer+2,SpeedNo
Speed_up2:      SJMP     STAR2
Speed_Down:     MOV      A,SpeedNo
                JZ       Speed_Down1
                DEC      SpeedNo
                MOV      buffer+2,SpeedNo
Speed_Down1:    SJMP     STAR2
Exec:           CLR      bFirst
                ACALL    TakeStepCount
                MOV      DPTR,#StepDelayTab
                MOV      A,SpeedNo
                MOVC     A,@A+DPTR
                MOV      StepDelay,A
                CJNE     A,#50,$+3
                JNC      Exec1
                MOV      A,#50
Exec1:          MOV      StartStepDelay,A
                MOV      StartStepDelay1,A
                SETB     TR0
                AJMP     STAR2
StepDelayTab:   DB       250,125,83,62,50,42,36,32,28,25,22,21
```

```
TIMER0:          PUSH     ACC
                 DJNZ     StartStepDelay,TIMER0_1
                 MOV      A,StartStepDelay1
                 CJNE     A,StepDelay,TIMER0_5
                 SJMP     TIMER0_2
TIMER0_5:        DEC      A
                 MOV      StartStepDelay1,A
TIMER0_2:        MOV      StartStepDelay,A
                 MOV      A,StepControl
                 CPL      A
                 MOV      P1,A
                 CPL      A
                 JB       bClockwise,TIMER0_3
                 RR       A
                 SJMP     TIMER0_4
TIMER0_3:        RL       A
TIMER0_4:        MOV      StepControl,A
                 MOV      A,R6
                 ORL      A,R7
                 JZ       TIMER0_1
                 SETB     bNeedDisplay
                 DJNZ     R7,TIMER0_1
                 DJNZ     R6,TIMER0_1
                 CLR      TR0
TIMER0_1:        POP      ACC
                 RETI
Step_SUB_1:      MOV      R5,#4
                 MOV      R0,#buffer+7
Step_SUB_1_1:    MOV      A,@R0
                 DEC      @R0
                 JNZ      Step_SUB_1_2
                 MOV      @R0,#9
                 DEC      R0
                 DJNZ     R5,Step_SUB_1_1
Step_SUB_1_2:    RET
TakeStepCount:   MOV      A,buffer+4         ;转动步数送入R6R7
                 MOV      B,#10
                 MUL      AB
                 ADD      A,buffer+5
                 MOV      B,#10
                 MUL      AB
                 ADD      A,buffer+6
                 MOV      R7,A
```

```
                    MOV     A,B
                    ADDC    A,#0
                    MOV     B,#10
                    MUL     AB
                    XCH     A,R7
                    MOV     B,#10
                    MUL     AB
                    XCH     A,B
                    ADD     A,R7
                    XCH     A,B
                    ADD     A,buffer+7
                    MOV     R7,A
                    MOV     A,B
                    ADDC    A,#0
                    MOV     R6,A
                    CJNE    R7,#0,TakeStepCount1
                    RET
TakeStepCount1:     INC     R6                      ;低位不为0,则高位加1,因循环时,会多减1
                    RET
                    END
```

八、实验扩展及思考

1. 怎样改变电机的转速？
2. 通过实验找出电机转速的上限，如何进一步提高最大转速？

实验 8　直流电机测速实验

一、实验目的

了解直流电机工作原理；了解光电开关的原理；掌握使用光电开关测量直流电机转速。

二、实验设备

STAR 系列实验仪一套、PC 机一台。

三、实验内容

1. 转速测量原理

本转速测量实验采用反射式光电开关，通过计数转盘通断光电开关产生的脉冲，计算出转速。

① 反射式光开关工作原理：光电开关发射光，射到测量物体上，如果强反射，光电开关接收到反射回来的光，则产生高电平 1，如图 4-8 所示；弱反射，光电开关接收不到反射回来的光，则产生弱电平 0，如图 4-9 所示。

②实验方法:本实验转速测量用的转盘在下表面做成如图 4-10 的转盘,白色部分为强反射区,黑色部分为弱反射区,转盘每转一圈,产生 4 个脉冲,每 1/4 秒计数出脉冲数,即得到每秒的转速。(演示程序中,LED 显示的是每秒钟转速)

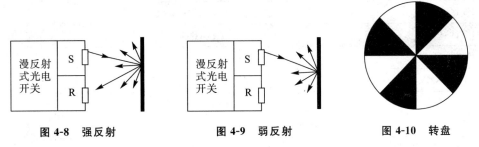

图 4-8　强反射　　　　　图 4-9　弱反射　　　　　图 4-10　转盘

2. 实验过程

①由 DAC0832 输出电压经功率放大后给电机供电,使用光电开关,测量电机转速,再经调整,最终将转速显示在 LED 上。

②通过按键调节电机转速,随之变化的转速动态显示在 LED 上。

四、实验原理图

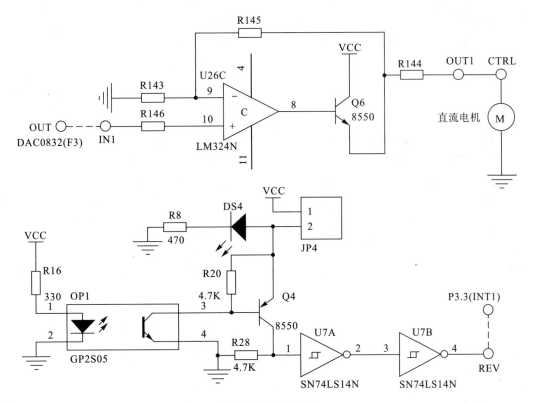

图 4-11　转速测量

五、实验步骤

1. 主机连线说明：

F3 区:CS	——	A3 区:CS1
F3 区:OUT	——	E2 区:IN1
E2 区:OUT1	——	F1:CTRL
F1 区:REV	——	A3 区:P3.3(INT1)
E5 区:CS、A0	——	A3 区:CS5、A0
E5 区:CLK	——	B2 区:2M
E5 区:A、B、C、D	——	G5 区:A、B、C、D

2. 由 DAC0832 输出电压经功率放大后驱动直流电机，通过单片机的计数器，计数光电开关通关次数并经过换算得出直流电机的转速，并将转速显示在 LED 上。

3. G5 区的 0、1 号按键控制直流电机转速快慢（最大转速≈96r/s,5V,误差±1r/s）。

六、演示程序

```
VoltageOffset   EQU     5               ;0832调整幅度
buffer          DATA    30H             ;需要8个字节的显示缓冲器
VOLTAGE         DATA    38H             ;转换电压数字量
Count           DATA    3AH             ;一秒转动次数
NowCountL       DATA    3BH             ;计数
NowCountH       DATA    3CH
kpTL1           DATA    3DH             ;保存上一次定时器1的值
kpTH1           DATA    3EH
DAC0832AD       XDATA   0F000H          ;DAC0832片选地址

                EXTRN   CODE(GetKeyA,Display8)

CSEG            AT      0000H           ;定位0
                LJMP    START
CSEG            AT      000BH           ;用于定时
                LJMP    TIME0
CSEG            AT      0013H
                LJMP    iINT1           ;光电开关产生脉冲,触发中断
CSEG            AT      0100H
START:          MOV     SP,#60H
                LCALL   MainINIT        ;初始化
MAIN:           LCALL   GetKeyA         ;按键扫描
                JNC     Main1
                JNZ     Key1
Key0:           MOV     A,#VoltageOffset ;0号键按下,转速提高
```

```
                ADD      A,VOLTAGE
                CJNE     A,VOLTAGE,$+3
                JNC      Key0_1
                MOV      A,#0FFH              ;最大
Key0_1:         MOV      VOLTAGE,A
                LCALL    DAC                  ;D/A
                SJMP     Main1
Key1:           MOV      A,VOLTAGE            ;1号键按下,转速降低
                CLR      C
                SUBB     A,#VoltageOffset
                JNC      Key1_1
                CLR      A                    ;最小
Key1_1:         MOV      VOLTAGE,A
                LCALL    DAC                  ;D/A
Main1:          JNB      F0,MAIN              ;F0=1,定时标志,刷新转速
                CLR      F0
                LCALL    RateTest             ;计算转速/显示
                JMP      MAIN                 ;循环进行实验内容介绍与测速功能测试
;主程序初始化
MainINIT:       CLR      F0                   ;请读取转速标志
                MOV      VOLTAGE,#99H         ;初始化转换电压输入值,99H—3.0V
                MOV      A,VOLTAGE
                LCALL    DAC                  ;初始D/A
;定时器/计数器初始化
                MOV      TMOD,#11H            ;开定时器0:定时方式1,定时器1:定时方式1
                MOV      R4,#5*4              ;定时5×50×4ms
                MOV      TL0,#0D4H            ;初始化定时器0,定时50ms(11.0592MHz)
                MOV      TH0,#4BH
                MOV      TL1,#00H             ;初始化器定时1
                MOV      TH1,#00H
                MOV      kpTL1,#00H           ;保存上一次定时器1的值
                MOV      kpTH1,#00H
                MOV      NowCountL,#0         ;计数器
                MOV      NowCountH,#0
                SETB     TR0                  ;开始定时
                SETB     TR1                  ;开始定时
                SETB     ET0                  ;开定时器0中断
                SETB     EX1                  ;开外部中断1
                SETB     IT1                  ;边沿触发
                SETB     EA                   ;允许中断
                RET
;定时器0中断服务程序
TIME0:          PUSH     ACC
```

```
            MOV     TL0,#0D5H        ;产生0.25s的定时(采用晶振11.0592MHz)
            MOV     TH0,#4BH
            DJNZ    R4,TIMER0_1
            SETB    F0               ;0.25*4s间隔标志F0
            MOV     R4,#5*4
            MOV     A,NowCountL
            RR      A
            RR      A
            ANL     A,#3FH
            MOV     Count,A
            MOV     A,NowCountH
            RR      A
            RR      A
            ANL     A,#0C0H
            ORL     Count,A          ;转一圈,产生4个脉冲,Count=NowCount/4
            MOV     NowCountL,#0
            MOV     NowCountH,#0
TIMER0_1:   POP     ACC
            RETI
iINT1:      PUSH    PSW              ;光电开关产生脉冲,触发中断
            PUSH    ACC
            CLR     TR1
            MOV     A,TL1
            CLR     C
            SUBB    A,kpTL1
            MOV     kpTL1,A
            MOV     A,TH1
            SUBB    A,kpTH1
            JNZ     iINT1_1
            MOV     A,kpTL1
            CJNE    A,#30H,$+3
            JC      iINT1_2          ;过滤干扰脉冲
iINT1_1:    INC     NowCountL
            MOV     A,NowCountL
            JNZ     iINT1_3
            INC     NowCountH
iINT1_3:    MOV     kpTL1,TL1
iINT1_2:    MOV     kpTH1,TH1
            SETB    TR1
            POP     ACC
            POP     PSW
            RETI
;转速测量/显示
```

```
RateTest:   MOV     A,Count
            MOV     B,#10
            DIV     AB
            JNZ     RateTest1
            MOV     A,#10H          ;高位为0,不需要显示
RateTest1:  MOV     buffer+1,A
            MOV     buffer,B
            MOV     A,VOLTAGE       ;给0832送的数据
            ANL     A,#0FH
            MOV     buffer+4,A
            MOV     A,VOLTAGE
            ANL     A,#0F0H
            SWAP    A
            MOV     buffer+5,A
            MOV     buffer+2,#10H   ;不显示
            MOV     buffer+3,#10H
            MOV     buffer+6,#10H
            MOV     buffer+7,#10H
            MOV     R0,#buffer
            LCALL   Display8        ;显示转换结果
            RET
;数模转换,A—转换数字量
DAC:        MOV     DPTR,#DAC0832AD
            MOVX    @DPTR,A
            RET
            END
```

七、实验扩展及思考题

在日光灯或白炽灯下,将转速分别调节到 25、50、75,观察转盘的变化。

实验 9　ISD1420 语音模块实验

一、实验目的

了解 ISD1420 的性能与单片机的接口逻辑;掌握手动和 MCU 控制两种录音、放音的基本功能。

二、实验设备

STAR 系列实验仪一套、PC 机一台。

三、实验内容

1. ISD1420 语音模块(B1 区)

①20 秒录放音长度,具有不掉电存储功能。

②分 1～160 段录放音片段。

2. 具体操作

①手动控制方式,通过 B1 区按键 REC 录音和按键 PLAYE、PLAYL 放音。

②MCU 控制方式,通过 G6 区 8 个按键控制录、放音:1～4 号键录音各 5 秒;然后通过 5～8 号键放音,放音内容顺序对应 1～4 号键的录音内容。

四、实验原理图

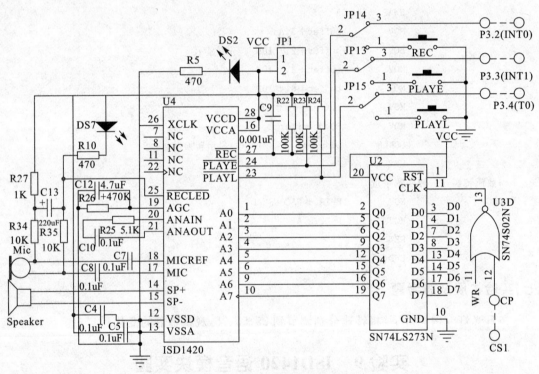

图 4-12 ISD1420 接口

五、实验步骤

1. 主机连线说明:

STAR ES598PCI

B1 区:REC	——	A3 区:P3.2(INT0) 录音控制
B1 区:PLAYE	——	A3 区:P3.3(INT1) 电平放音控制
B1 区:PLAYL	——	A3 区:P3.4(T0) 触发放音控制,下降沿触发
B1 区:CP	——	A3 区:CS1
G6 区:JP74	——	A3 区:JP51(P1 口)

STAR ES59PA

B1 区：REC	——	A3 区：P3.2(INT0) 录音控制
B1 区：PLAYE	——	A3 区：P3.3(INT1) 电平放音控制
B1 区：PLAYL	——	A3 区：P3.4(T0) 触发放音控制，下降沿触发
B1 区：JP107	——	B3 区：JP66
B3 区：CS_273	——	A3 区：CS1
G6 区：JP74	——	A3 区：JP51(P1 口)

2. 将 JP13、JP14、JP15 跳向"MANUAL"，即手动录、放音。3 个控制按键(在 B1 区左边)REC(录音)、PLAYE(电平放音)和 PLAYL(边沿放音)控制录音和放音。

3. 将 JP13、JP14、JP15 跳向"MCU"，单片机控制，运行演示程序，1～4 号键录音，5～8 号键放音。

六、演示程序

1. ISD1420 子程序(ISD1420.ASM)

```
REC         BIT     P3.2                ;录音接口
PLAYE       BIT     P3.3                ;电平触发放音接口
PLAYL       BIT     P3.4                ;边沿触发放音接口
ISDCOMM     XDATA   0F000H              ;录放音地址/操作模式输入地址
ISD_INIT:   SETB    REC                 ;语音模块初始化,关闭录放音功能
            SETB    PLAYE
            SETB    PLAYL
            MOV     DPTR,#ISDCOMM
            CLR     A
            MOVX    @DPTR,A             ;允许手动操作,当A6,A7为高时,无法手动操作
            RET
;操作模式,A－操作模式设置值
ISD_MODE:   PUSH    ACC
            LCALL   ISD_STOP            ;语音初始化,置位 REC,PLAYE,PLAYL,并设置操作模式
            MOV     DPTR,#ISDCOMM       ;设置操作模式:分段录音
            POP     ACC
            MOVX    @DPTR,A             ;设置操作模式命令(在 A 中)
            CLR     PLAYL               ;给一个上升沿,锁存命令
            NOP
            NOP
            NOP
            SETB    PLAYL
            RET
;录音
ISD_REC:    MOV     DPTR,#ISDCOMM       ;设置录音起始地址
            MOVX    @DPTR,A
            CLR     REC                 ;REC 变低,即开始录音
```

```
                        RET
;放音子程序
;A:放哪段音
ISD_PLAY:       PUSH    ACC
                CALL    ISD_STOP                ;暂停之前的录放音操作
                POP     ACC
                MOV     DPTR,#ISDCOMM           ;设置放音起始地址
                MOVX    @DPTR,A
                CLR     PLAYE                   ;开始放音,边沿放音模式
                NOP
                SETB    PLAYE
                RET
;停止录放音
ISD_STOP:       CLR     PLAYL                   ;一个负脉冲停止放音
                NOP
                SETB    PLAYL
                LCALL   Delay50ms
                SETB    REC                     ;关闭所有操作指令
                SETB    PLAYE
                MOV     DPTR,#ISDCOMM
                CLR     A
                MOVX    @DPTR,A                 ;允许手动录放音,当A6,A7为高时,无法手动放音
                RET
```

2. 主程序(MAIN.ASM)

```
ISD1420_AD1     EQU     00H                     ;1号键录放音起始地址,每次录音5s
ISD1420_AD2     EQU     28H                     ;2号键录放音起始地址
ISD1420_AD3     EQU     50H                     ;3号键录放音起始地址
ISD1420_AD4     EQU     78H                     ;4号键录放音起始地址
```

(1)录放音子程序

```
KEY1:           MOV     A,#ISD1420_AD1          ;录音首地址
                LJMP    KEY_REC
KEY2:           MOV     A,#ISD1420_AD2
                LJMP    KEY_REC
KEY3:           MOV     A,#ISD1420_AD3
                LJMP    KEY_REC
KEY4:           MOV     A,#ISD1420_AD4
                LJMP    KEY_REC
KEY_REC:        MOV     R7,#20                  ;录音时间长度,5s
                LCALL   ISD_REC                 ;调用录音子程序
KEY_REC1:       LCALL   Delay_025S              ;延时
                JB      F0,KEY_REC2             ;检测按键是否有键按下
                DJNZ    R7,KEY_REC1             ;录音时间,根据R7的值决定
                LCALL   ISD_STOP                ;停止录音
```

```
        KEY_REC2:    RET
```
(2)放音子程序
```
        KEY5:        MOV      A,#ISD1420_AD1      ;放音首地址
                     LJMP     KEY_PLAY
        KEY6:        MOV      A,#ISD1420_AD2
                     LJMP     KEY_PLAY
        KEY7:        MOV      A,#ISD1420_AD3
                     LJMP     KEY_PLAY
        KEY8:        MOV      A,#ISD1420_AD4
                     LJMP     KEY_PLAY
        KEY_PLAY:    MOV      R7,#20
                     LCALL    ISD_PLAY            ;调用录音子程序
        KEY_PLAY1:   LCALL    Delay_025S          ;用于进度显示的时间参照
                     JB       F0,KEY_PLAY2        ;检测按键是否有键按下
                     DJNZ     R7,KEY_PLAY1
        KEY_PLAY2:   RET
```

七、实验扩展及思考题

实验名称：公交车的报站功能。

实验内容：利用掌握分段录音和放音控制，实现公交车的报站功能，有兴趣者可自行尝试。

实验 10 CAN 通信实验

一、实验目的

1. 了解 CAN 总线工作原理。

2. 掌握使用 SJA1000 进行 CAN 总线通讯的方法。

二、实验设备

STAR 系列实验仪一套、PC 机一台、CAN2.0(SJA1000)模块两块。

三、实验内容

1. 熟悉 SJA1000。

①与 CAN2.0B 协议兼容；二种工作模式：BasicCAN、PeliCAN。

②位速率可达 1Mbit/s；64 字节 FIFO。

③支持热插拔；可替换 PCA82C200。

AD0～7:地址/数据总线　　　　　　　　ALE:地址锁存使能端

CS:片选　　　　　　　　　　　　　　CLKOUT:可编程时钟输出

RD:读选通　　　　　　　　　　　　　WR:写选通

XTAL1、XTAL2:振荡频率输入　　　　　MODE:1 表示 Intel;0 表示 Motolora

$V_{DD}3$、$V_{SS}3$：驱动输出的电压源　　　　　$V_{DD}2$、$V_{SS}2$：比较器电压源
TX0：CAN 输出 0　　　　　　　　　　　　　TX1：CAN 输出 1
RX0、RX1：CAN 输入，输入到 CAN 的输入比较器　INT：中断，低电平有效
RST：复位，低电平有效　　　　　　　　　　　$V_{DD}1$、$V_{ss}1$：逻辑电路的电压源

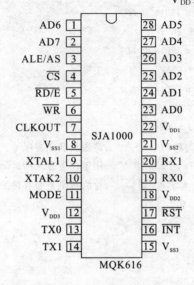

图 4-13　SJA1000 引脚图

2. 熟悉、理解 SJA1000 二种工作模式：BasicCAN、PeliCAN。
3. 熟悉、理解程序库 star51.lib 的使用，编写 CAN 通信程序。
4. 通过模块 1 向模块 2 循环发送 16K 数据，如果正确接收，通过 P1.7，点亮 G6 区的指示灯。

四、实验步骤

1. 连线说明：

CAN 模块 1	——	B5 区
A3 区：INT0	——	B5 区：CZ1-1(SJA1000 INT)
A3 区：CS1	——	B5 区：CZ1-2(SJA1000 CS)
A3 区：P1.1	——	B5 区：CZ1-3(TJA1050 S)
A3 区：P1.0	——	B5 区：CS1-7(SJA1000 RST)
CAN 模块 2	——	C6 区
A3 区：INT1	——	C6 区：CZ2-1(SJA1000 INT)
A3 区：CS2	——	C6 区：CZ2-2(SJA1000 CS)
A3 区：P1.3	——	C6 区：CZ2-3(TJA1050 S)
A3 区：P1.2	——	C6 区：CZ2-7(SJA1000 RST)
A3 区：JP51	——	G6 区：JP65

2. 使用星研集成环境软件,打开 CAN 中项目(CAN.PRV)。

3. 编译、连接后,运行,传送正确,P1.7 对应的指示灯亮;传送错误,P1.7 对应的指示灯熄灭。

五、程序示例

```
#include "reg52.h"
#include "SJA1000.h"                       //SJA1000 常量、数据结构的定义
#include "can.h"                           //库文件中有关 CAN 通信的函数、变量说明

sbit ResetCan1=P1^0;                       //复位 CAN1
sbit S_Can1=P1^1;                          //选择高速通信
sbit ResetCan2=P1^2;                       //复位 CAN2
sbit S_Can2=P1^3;                          //选择高速通信
sbit bCanOK=P1^7;                          //0:通信正确,1:错误
xdata uchar can_cs1 _at_ 0xf000;           //CAN1 基地址
xdata uchar can_cs2 _at_ 0xe000;           //CAN2 基地址
#if defined (PeliCANMode)                  //PeliCAN mode
code const uchar can1_FrameInfo_ACR[5]={SFF_FRAMEINFO, ClrByte, ClrByte, ClrByte, ClrByte};
code const uchar can2_FrameInfo_ACR[5]={SFF_FRAMEINFO, ClrByte, ClrByte, ClrByte, ClrByte};
code const uchar can1_AMR[4]={DontCare, DontCare, DontCare, DontCare};
code const uchar can2_AMR[4]={DontCare, DontCare, DontCare, DontCare};
#else                                      //BasicCAN mode
code const uchar can1_FrameInfo_ACR[1]={ClrByte};
code const uchar can2_FrameInfo_ACR[1]={ClrByte};
code const uchar can1_AMR[1]={DontCare};
code const uchar can2_AMR[1]={DontCare};
#endif

struct TRANSMIT_STRUCT transmit_struct, *pTransmit;   //发送结构
struct RECEIVE_STRUCT receive_struct;                 //接收结构
xdata uchar TransmitBuffer[0x4000] _at_ 0x2000;       //16K 发送数据区
void delay()                                          //复位延时
{
    int i;
    for(i=0; i<0x7fff; i++)
    {
    }
}
main()
{
    int length, length1;
    bit bTRUE;                                        //接收正确标志
```

```
        uchar cLength;                           //接收一帧数据的长度

    ResetCan1=0;                             //复位
    ResetCan2=0;
    delay();
    ResetCan1=1;
    ResetCan2=1;
    S_Can1=0;                                //高速
    S_Can2=0;

//初始化 SJA1000
#if defined (PeliCANMode)
    pCS_SJA1000=&can_cs1;                    //初始化 CAN1

    init_SJA1000(
//CDR 选择 PeliCAN Mode,旁路输入比较器,关闭 CLKOUT
            CANMode_Bit | CBP_Bit | ClkOff_Bit,
            can1_FrameInfo_ACR+1,
            can1_AMR,                        //every identifier is accepted
//bit-rate: 100 kBit/s ,oscillator frequency: 16 MHz,1,0 %
//maximum tolerated propagation delay: 4450 ns ,minimum requested propagation delay: 500 ns
            0x41,0x1c,
//configure CAN outputs: float on TX1, Push/Pull on TX0, normal output mode
            Tx1Float | Tx0PshPull | NormalMode);
    pCS_SJA1000=&can_cs2;                    //初始化 CAN2
    init_SJA1000(
//CDR 选择 PeliCAN Mode,旁路输入比较器,关闭 CLKOUT
            CANMode_Bit | CBP_Bit | ClkOff_Bit,
            can2_FrameInfo_ACR+1,
            can2_AMR,                        //every identifier is accepted
//bit-rate: 100 kBit/s ,oscillator frequency: 16 MHz,1,0 %
//maximum tolerated propagation delay: 4450 ns,minimum requested propagation delay:500 ns 0x41,0x1c,
//configure CAN outputs: float on TX1, Push/Pull on TX0, normal output mode Tx1Float | Tx0PshPull | NormalMode);
#else /* BasicCAN mode */
    pCS_SJA1000=&can_cs1;
    init_SJA1000(
//CDR 选择 BasicCAN Mode,旁路输入比较器,关闭 CLKOUT
            CBP_Bit | ClkOff_Bit,
            can1_FrameInfo_ACR,
            can1_AMR,                        //every identifier is accepted
//bit-rate: 100 kBit/s ,oscillator frequency: 16 MHz, 1,0 %
//maximum tolerated propagation delay: 4450 ns ,minimum requested propagation delay:500 ns 0x41,0x1c,
```

```
//configure CAN outputs: float on TX1, Push/Pull on TX0, normal output mode TxlFloat | Tx0PshPull | NormalMode);
    pCS_SJA1000=&can_cs2;
    init_SJA1000(
//CDR 选择 BasicCAN Mode,旁路输入比较器,关闭 CLKOUT
            CBP_Bit | ClkOff_Bit,
            can1_FrameInfo_ACR,
            can1_AMR,                          //every identifier is accepted
//bit-rate : 100 kBit/s ,oscillator frequency : 16 MHz, 1,0 %
//maximum tolerated propagation delay : 4450 ns ,minimum requested propagation delay:500 ns 0x41,0x1c,
//configure CAN outputs:float on TX1,Push/Pull on TX0,normal output mode TxlFloat | Tx0PshPull | NormalMode);
#endif

    EX0=1;                                     //外部中断 0:CAN1
    EX1=1;                                     //外部中断 1:CAN2
    PX0=1;
    EA=1;

    while (1)
    {
        receive_struct.startAdr=receive_struct.endAdr=0;
        //初始化接收结构,无数据启动发送过程
        transmit_struct.wLength=0x4000;        //发送 16K
        transmit_struct.wCount=0;
        transmit_struct.pData=TransmitBuffer;  //发送哪块
        transmit_struct.pID=can1_FrameInfo_ACR;
        pCS_SJA1000=&can_cs2;                  //第二个 CAN 向第一个 CAN 发数据
        pTransmit=&transmit_struct;
        transmit();                            //发送 pTransmit 指向的发送结构
        length=0;
        bTRUE=TRUE;
        bIncomplete=FALSE;                     //终止发送时,上次发送未完成
        bDataOverrun=FALSE;                    //数据溢出
        do
        {
            while (receive_struct.startAdr==receive_struct.endAdr);
                //是否接收到数据,接收结构中已接收数据长度
            length1=((receive_struct.endAdr-receive_struct.startAdr) &
                    MAX_CONTROLDATA_SIZE_AND);
            if (length1+100 > MAX_CONTROLDATA_SIZE_AND)
                EA=0;                          //已接收数据太多,关中断,不允许发送、接收
            else
                EA=1;                          //接收缓冲器允许接收数据
            if (bDataOverrun)
```

```c
            {
                bTRUE=FALSE;                    //SJA1000 接收 FIFO 溢出错
                break;
            }
#if defined (PeliCANMode)
            cLength=receive_struct.dataBuffer[receive_struct.startAdr++];
            if (cLength & 0x80)                 //不理睬识别码,起始指针移至数据区
                receive_struct.startAdr+=4;     //扩展帧
            else
                receive_struct.startAdr+=2;     //标准帧
#else /* BasicCAN mode */
            {
                receive_struct.startAdr++;      //不理睬识别码,起始指针移至数据区
                receive_struct.startAdr &=MAX_CONTROLDATA_SIZE_AND;
                cLength=receive_struct.dataBuffer[receive_struct.startAdr++];
            }
#endif
            cLength &=0xf;                      //一帧数据的长度
            receive_struct.startAdr &=MAX_CONTROLDATA_SIZE_AND;
            //保证循环接收队列不超界
            while (cLength--)
            {
                if(receive_struct.dataBuffer[receive_struct.startAdr++]!=
                    TransmitBuffer[length++])
                {//接收数据错误
                    bTRUE=FALSE;
                    break;
                }
                //保证循环接收队列不超界
                receive_struct.startAdr &=MAX_CONTROLDATA_SIZE_AND;
            }
            if (!bTRUE)
            break;
        }while (length < 0x4000);               //接收 16K
        if (bTRUE)
            bCanOK=0;                           //P1.7 指示灯亮,表示传送正确
        else
        {
            AbortTransmit();                    //错误,终止发送
            bCanOK=1;
        }
    }
}
```

```
void can_int_0 (void) interrupt 0
{
    uchar * pCS_SJA1000_1=pCS_SJA1000;
    pCS_SJA1000=&can_cs1;                    //接收数据
    SJA1000_INT();
    pCS_SJA1000=pCS_SJA1000_1;
}

void can_int_1 (void) interrupt 2
{
    uchar * pCS_SJA1000_1=pCS_SJA1000;
    pCS_SJA1000=&can_cs2;                    //第二个 CAN 向第一个 CAN 发数据
    SJA1000_INT();
    pCS_SJA1000=pCS_SJA1000_1;
}
```

实验 11　USB 2.0 通信实验

一、实验目的

1. 了解 USB 串行通讯工作原理。
2. 掌握实现 USB 串行通讯功能的方法。

二、实验设备

STAR 系列实验仪一套、PC 机一台、ISP1581(USB 2.0)模块一块。

三、实验内容

1. 熟悉 ISP1581。
① 完全符合通用串行总线(USB)Rev 2.0 规范。
② 内带高速 DMA,8K Bytes FIFO Memory。
③ 7 个 IN 端点、7 个 OUT 端点、1 个控制 IN/OUT 端点。
④ 每个端点都可以配置双缓冲器,轻松实现实时数据传输。
⑤ 内部集成了 PLL,配置 12MHz 晶振,良好的 EMI 特性。
⑥ 支持大部分微控制器/微处理器的总线接口。
⑦ 自动识别 USB 2.0、USB 1.1 模式。
⑧ 可通过软件控制与 USB 总线的连接(SoftConnect)。
⑨ 集成了 SIE、PIE、FIFO、5V 转 3.3V 的电压调整器。
⑩ 符合 ACPITM、OnNowTM 和 USB 电源管理的要求。
2. 熟悉、理解 ISP1581\ISP1581 中的所有固件程序。
3. 熟悉、理解 ISP1581\ISP1581_PC 部分中 STAR_ISP1581 驱动程序库的使用。

4.通过 USB 与 PC 进行数据传输,并检验传输数据的正误。

四、实验步骤

1.连线说明:

B5 区:KZ1-1(INT)	——	A3 区:P3.2(INT0)
B5 区:KZ1-2(CS)	——	A3 区:CS2
B5 区:KZ1-7(RESET_N)	——	A3 区:P1.6

将 USB 通信线一端与 ISP1581(USB2.0)模块相连,另一端插入电脑。

2.使用星研集成环境软件,打开 ISP1581\ISP1581 中固件项目(ISP1581.PRV)。

3.运行 ISP1581_Test.exe。分别使用扫描仪方式、打印机方式、循环读写方式测试通信稳定性。

六、程序示例

1. ISP1581 固件

ISP1581 固件例子在 ISP1581 目录中,根据需要可以修改 main()函数,调用 ReadEndpoint 读数据、WriteEndpoint 写数据,进行数据收发。注意更改 USB_Int_Flag 中标志前,必须先关中断,这是因为 USB 中断程序会修改 USB_Int_Flag。

2. ISP1581_Test.exe

对 ISP1581 目录中的程序编译、连接后传送到仿真器中,开始运行,系统提示找到 USB 设备,需要驱动程序:根据使用的操作系统,选择 Win98_Win2000 或 XP 目录。系统会自动装载驱动程序。

ISP1581.exe 是一个例子程序。执行后,缓冲区大小可选择 16384,然后可以使用扫描仪方式、打印机方式、循环读写方式测试。通用输出(端点 1)需要对 main()作一些调整,根据收到的 4 个数据(只有一个数据有效),作一些显示。

3. STAR_ISP1581 驱动程序库的使用

它包含 3 个文件:Star_ISP1581.dll、Star_ISP1581.lib、StarISP1581.h。动态链接库是 Microsoft Windows 的标准接口,流行的开发工具 VC、VB、VF、Delphi、C++Builder、PowerBuilder 等均可使用,可以选择自己熟悉的工具进行 USB 开发。它提供 4 个函数:

(1)读 USB 的 Endpoint1

BOOL _declspec(dllimport) ReadPort1(unsigned char * lpBuffer, int iNumber);

正确,返回 TRUE;错误,返回 FALSE。

例子:

```
unsigned char Buffer[100];
if (ReadPort1(Buffer,100))
{
}
```

(2)读 USB 的 Endpoint2

BOOL _declspec(dllimport) ReadPort2(unsigned char * lpBuffer, int iNumber);

正确,返回 TRUE;错误,返回 FALSE。
```
unsigned char Buffer[100];
if (ReadPort2(Buffer,100))
{
}
```

(3)写 USB 的 Endpoint1

BOOL _declspec(dllimport) WritePort1(unsigned char * lpBuffer, int iNumber);

正确,返回 TRUE;错误,返回 FALSE。
```
int i;
unsigned char Buffer[100];
for (i=0; i < 100; i++)
    Buffer[i]=i+1;
if (WritePort1(Buffer,100))
{
}
```

(4)写 USB 的 Endpoint2

BOOL _declspec(dllimport) WritePort2(unsigned char * lpBuffer, int iNumber);

正确,返回 TRUE;错误,返回 FALSE。
```
int i;
unsigned char Buffer[100];
for (i=0; i < 100; i++)
    Buffer[i]=i+1;
if (WritePort2(Buffer,100))
{
}
```

七、实验扩展及思考题

1. 使用 ISP1581(USB2.0)模块,设计一个虚拟示波器。
2. 使用 ISP1581(USB2.0)模块,控制实验仪的其他模块。

实验 12 触摸屏实验(ADS7843、12864C)

一、实验目的

1. 了解图形液晶模块的控制方法;了解它与单片机的接口逻辑;掌握使用图形点阵液晶显示字体和图形。

2. 了解触摸屏结构、工作原理;掌握触摸屏控制芯片 ADS7843 的使用方法。

二、实验设备

STAR 系列实验仪一套、PC 机一台。

三、实验资料简介

1. 12864C 液晶显示器

12864C 液晶与 12864J 使用相同的控制芯片，它们的引脚排列不同，它们的控制程序完全相同。

2. 触摸屏的基本原理

典型触摸屏的工作部分由三部分组成，如图 4-14 所示：两层透明的阻性导体层、两层导体之间的隔离层、电极。阻性导体层选用阻性材料，如铟锡氧化物（ITO）涂在衬底上，上层衬底用塑料，下层衬底用玻璃。隔离层为黏性绝缘液体材料，如聚酯薄膜。电极选用导电性能极好的材料（如银粉墨），其导电性能大约为 ITO 的 1000 倍。

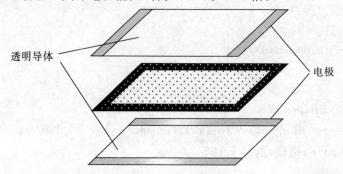

图 4-14　触摸屏结构

触摸屏工作时，上下导体层相当于电阻网络，如图 4-15 所示。当某一层电极加上电压时，会在该网络上形成电压梯度。如有外力使得上下两层在某一点接触，则在电极未加电压的另一层可以测得接触点处的电压，从而知道接触点处的坐标。比如，在顶层的电极（X＋，X－）上加上电压，则在顶层导体层上形成电压梯度，当有外力使得上下两层在某一点接触，在底层就可以测得接触点处的电压，再根据该电压与电极（X＋）之间的距离关系，知道该处的 X 坐标。然后，将电压切换到底层电极（Y＋，Y－）上，并在顶层测量接触点处的电压，从而知道 Y 坐标。

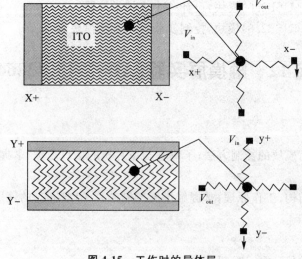

图 4-15　工作时的导体层

3. 触摸屏控制芯片 ADS7843

4 线触摸屏控制芯片内置 12 位模数转换,可选择 8 位、12 位模式工作;供电电压 2.7～5V;最高转换速率为 125kHz;低导通电阻模拟开关的串行接口芯片;参考电压 VREF 为 1V～+VCC,转换电压的输入范围为 0～VREF。

四、实验内容

1. 在液晶屏上显示几幅画面。
2. 每幅画面上,均有 1～4 个带有椭圆边框的有效触摸区域,相当于按键。
3. 在程序中,每幅画面中建立每个有效触摸区域的矩形坐标。
4. 根据画面编号、矩形坐标,识别每个有效触摸区域,转化为键值。
5. 根据键值,切换画面。

五、实验原理图

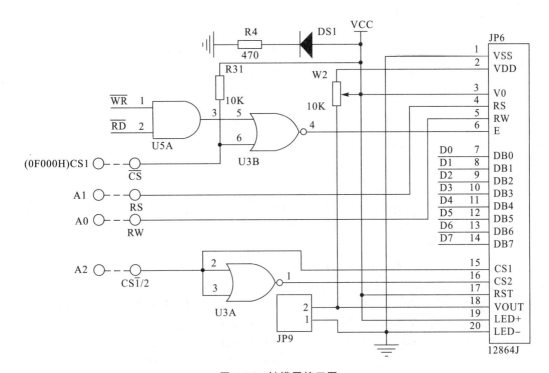

图 4-16 触摸屏接口图

六、实验步骤

1. 主机连线说明:

A1 区:CS,RW,RS,CS1/2	——	A3 区:CS1、A0、A1、A2
A1 区:CS,DCLK,DIN,DOUT,INT	——	A3 区:P1.1、P1.0、P1.2、P1.4、INT1

2. 运行程序,验证显示结果。

七、演示程序

1. 整屏显示子程序。

```
Draw_A_Picture:  MOV    A,#0         ;起始行,第 0 行
                 MOV    B,#0         ;起始列,第 0 列
DrawL1:          PUSH   ACC
                 MOV    R7,#64       ;一行共占 8*8 个字节
                 LCALL  SETXYL       ;设置起始显示行列地址(左边屏)
DrawL2:          CLR    A
                 MOVC   A,@A+DPTR
                 INC    DPTR
                 LCALL  WRDATAL
                 DJNZ   R7,DrawL2
                 POP    ACC
DrawR1:          PUSH   ACC
                 MOV    R7,#64
                 LCALL  SETXYR       ;设置起始显示行列地址(右边屏)
DrawR2:          CLR    A
                 MOVC   A,@A+DPTR
                 INC    DPTR
                 LCALL  WRDATAR
                 DJNZ   R7,DrawR2
                 POP    ACC
                 INC    A
                 CJNE   A,#8,DrawL1  ;共 8 行
                 RET
```

2. ADS7843 子程序。

```
                 NAME   ADS7843
;触摸屏控制芯片 ADS7843 控制字
X_8_Com          EQU    9CH          ;控制字 选择 X 通道,8 位精度,参考电压:差动模式
Y_8_Com          EQU    0DCH         ;控制字 选择 Y 通道,8 位精度,参考电压:差动模式
X_12_Com         EQU    94H          ;控制字 选择 X 通道,12 位精度,参考电压:差动模式
Y_12_Com         EQU    0D4H         ;控制字 选择 Y 通道,12 位精度,参考电压:差动模式
#ifdef           bADS7843_12bit
PUBLIC           xCoordinateL,xCoordinateH,yCoordinateL,yCoordinateH
PUBLIC           key,ReadCoordinate
#else
PUBLIC           xCoordinate, yCoordinate
PUBLIC           key,ReadCoordinate
#endif
ADS7843_CODE     SEGMENT CODE
ADS7843_DATA     SEGMENT DATA
```

```
;控制管脚
CLK         BIT     P1.0            ;串行时钟
CS          BIT     P1.1            ;片选
DIN         BIT     P1.2            ;串行输入
DOUT        BIT     P1.4            ;串行输出
bBusy       BIT     P1.3            ;查忙标志
;数据段
RSEG        ADS7843_DATA
#ifdef bADS7843_12bit
xCoordinateL:   DS      1           ;触摸点 X 方向坐标(0-0FFFH)(低 8 位)
xCoordinateH:   DS      1           ;触摸点 X 方向坐标(0-0FFFH)(高 4 位)
yCoordinateL:   DS      1           ;触摸点 Y 方向坐标(0-0FFFH)(低 8 位)
yCoordinateH:   DS      1           ;触摸点 Y 方向坐标(0-0FFFH)(高 4 位)
#else
xCoordinate:    DS      1           ;触摸点 X 方向坐标(0-0FFH)
yCoordinate:    DS      1           ;触摸点 Y 方向坐标(0-0FFH)
#endif
;代码段
RSEG        ADS7843_CODE
;读取坐标,CY=0,正确;CY=1,失败
ReadCoordinate:                     ;取坐标
            SETB    CS              ;产生开始信号
            CLR     CLK
            SETB    DIN
            CLR     CS
#ifdef bADS7843_12bit
            MOV     A,#X_12_Com
            CALL    Write_8_bits
            CALL    T_Busy
            JC      ReadCoordinate_Exit
            CALL    Read_12_bits
            MOV     xCoordinateH,A
            MOV     xCoordinateL,B
            MOV     A,#Y_12_Com
            CALL    Write_8_bits
            CALL    T_Busy
            JC      ReadCoordinate_Exit
            CALL    Read_12_bits
            MOV     yCoordinateH,A
            MOV     yCoordinateL,B
#else
            MOV     A,#X_8_Com
            CALL    Write_8_bits
```

```
                CALL    T_Busy
                JC      ReadCoordinate_Exit
                CALL    Read_8_bits
                MOV     xCoordinate,A
                MOV     A,#Y_8_Com
                CALL    Write_8_bits
                CALL    T_Busy
                JC      ReadCoordinate_Exit
                CALL    Read_8_bits
                MOV     yCoordinate,A
#endif
                CLR     C
ReadCoordinate_Exit:
                RET
;查忙状态,CY=0,成功;CY=1,失败
T_Busy:         PUSH    07H
                MOV     R7,#50
                CLR     C
T_Busy1:        JB      bBusy,T_Busy_Exit
                DJNZ    R7,T_Busy1
                SETB    C
T_Busy_Exit:    POP     07H
                RET
;写8位数据(参数在 ACC 里)
Write_8_bits:   PUSH    07H
                MOV     R7,#8
W_8_b:          RLC     A
                MOV     DIN,C
                SETB    CLK
                NOP
                CLR     CLK
                DJNZ    R7,W_8_b
                POP     07H
                RET
;读取8位数据(结果存放于在 ACC 里)
Read_8_bits:    PUSH    07H
                MOV     R7,#8
R_8_b:          SETB    CLK
                NOP
                CLR     CLK
                MOV     C,DOUT
                RLC     A
                DJNZ    R7,R_8_b
```

```
                POP     07H
                RET
;读取12位数据(结果存放于在A(高位)、B里)
Read_12_bits:   CALL    Read_8_bits
                MOV     B,A
                CALL    Read_8_bits
                ANL     A,#0F0H
                XCH     A,B
                PUSH    ACC
                ANL     A,#0FH
                ORL     A,B
                SWAP    A
                MOV     B,A
                POP     ACC
                ANL     A,#0F0H
                SWAP    A
                RET
;DPTR指向的第一个参数是键数(1个字节),第二个参数开始:键的坐标点(4个字节x1,x2,y1,y2),
;返回键值在A中,如果A=0FFH,表示没有对应按键
key:            PUSH    07H
                PUSH    B
                CALL    ReadCoordinate
                JC      NoKey
                CALL    Delay20ms       ;延时去抖动
                CALL    ReadCoordinate
                JC      NoKey
                MOV     B,#0
                CLR     A
                MOVC    A,@A+DPTR
                MOV     R7,A
                INC     DPTR
key1:           CLR     A
                MOVC    A,@A+DPTR
                INC     DPTR
                CJNE    A,xCoordinate,$+3
                JNC     key1_1
                CLR     A
                MOVC    A,@A+DPTR
                INC     DPTR
                CJNE    A,xCoordinate,$+3
                JC      key1_2
                CLR     A
                MOVC    A,@A+DPTR
```

```
                INC    DPTR
                CJNE   A,yCoordinate,$+3
                JNC    key1_3
                CLR    A
                MOVC   A,@A+DPTR
                INC    DPTR
                CJNE   A,yCoordinate,$+3
                JNC    key_exit
key1_4:         INC    B
                DJNZ   R7,key1
NoKey:          MOV    B,#0FFH
key_exit:       MOV    A,B
                POP    B
                POP    07H
                JNB    P3.3,$
                CALL   Delay20ms
                CLR    IT1
                SETB   EX1
                RET
key1_1:         INC    DPTR
key1_2:         INC    DPTR
key1_3:         INC    DPTR
                SJMP   key1_4

Delay20ms:      MOV    R6,#40        ;延时 20ms
                MOV    R7,#0
Delay20ms_1:    DJNZ   R7,$
                DJNZ   R6,Delay20ms_1
                RET
```

实验 13　GPS 定位实验

一、实验目的

1. 了解 GPS 模块的各项参数，能够选择合适的场合使用 GPS 模块。
2. 了解 NMEA-0183 标准输出格式。
3. 掌握使用 GPS 模块，在 GPRMC 中提取日期、时间、纬度、经度、海拔。

二、实验设备

STAR 系列实验仪一套、PC 机一台、GPS 模块一块。

三、实验内容

1. GPS 模块

本 GPS 模块使用内置天线,应置于室外或窗口,否则,内置天线无法接收卫星信号,GPS 模块无法正常工作。调节 GPS 模块位置、方向,使它追踪尽可能多的卫星,测得的数据就更正确。

GPS 模块缺省的通信方式:波特率 4800bps、1 位起始位、8 位数据位、1 位停止位。

使用 NMEA-0183 2.4 版、ASCII 输出标准输出格式,可选用 SiRF 二进制格式。

开机时,GPS 模块开始定位,正常状况下,定位约需 45 秒。定位后,输出有效的经度、纬度、高度、速度、日期、时间、估计误差值、卫星状态、接收状态。

NMEA-0183 标准输出格式:

种 类	说 明
GPGGA	卫星定位资讯(指定位后)
GPGLL	地理位置——经度及纬度
GPGSA	GNSS DOP(一种偏差资讯,说明卫星定位讯号的优劣状态)
GPGSV	GNSS 天空范围内的卫星
GPRMC	最起码的 GNSS 资讯(指达到定位目的)
GPVTG	对地方向及对地速度

(1)卫星定位资讯(GPGGA)

输出范例:

信号代号	$ GPGGA	GPGGA 规范抬头
标准定位时间	161229.487	时时分分秒秒.秒秒秒
纬度	3723.2475	度度分分.分分分分
北半球或南半球指示器	N	北半球(N)或南半球(S)
经度	12158.3416	度度度分分.分分分分
东半球或西半球指示器	W	东(E)半球或西(W)半球
定位代号指示器	1	0:未定位或无效的定位
		1:GPSSPS 格式(商业用途格式),已定位
		2:偏差修正 GPS(即 DGPS),SPS 格式,已定位
		3:GPS PPS 格式(PPS 为军用格式),已定位
使用中的卫星数目	07	00 至 12
水平稀释精度	1.0	0.5 至 99.9 米
海拔高度	9.0 米	−9999.9 至 99999.9 米
单位	m	米
地表平均高度	米	−999.9 至 9999.9 米
单位	m	米
差分修正 DGPS		RTCM SC−104)资料年限,上次有效的 RTCM 传输至今的秒数(若非 DGPS,则数字为 0)
偏差修正(DGPS)		参考基地台代号,0000 至 1023。(0 表非 DGPS)

差分参考基站代码 ID	0000	
总和校验码	*18	
<CR> <LF>		讯息终点

(2) 最起码的 GNSS 资讯(GPRMC)

输出范例：

信号代号	$ GPRMC	GPRMC 规范抬头
标准定位时间	161229.487	时时分分秒秒.秒秒秒
定位状态	A	A=资料可用，V=资料不可用
纬度	3723.2475	度度分分.分分分分
北半球或南半球指示器	N	北半球(N)或南半球(S)
经度	12158.3416	度度度分分.分分分分
东半球或西半球指示器	W	东(E)半球或西(W)半球
对地速度	0.13　节	0.0 至 1851.8 节
对地方向	309.62　度	实际值
日期	120598	日日月月年年
磁极变量(A)	度	东(E)半球或西(W)半球
总和校验码	*10	
<CR> <LF>		讯息终点

SiRF 公司目前不支持磁极变量，所有对地方向资料是以大地测量 WGS84 为方向。

2. 编写程序

编程：通过 GPGGA、GPRMC，提取日期、时间、纬度、经度、海拔，并显示。

四、实验步骤

1. 连线说明：

GPS 专用转接线的一端(带有红、蓝连接线)	——	E7 区：J1A3 区
红线	——	C1 区：VCC
蓝线	——	C1 区：GND
GPS 专用转接线的另一端		GPS 模块
A3 区：RXD、TXD	——	E7 区：RXD、TXD
A3 区：CS1、A0、A1、A2	——	A1 区：CS、RW、RS、CS1/2

2. 编写、调试程序，在液晶上显示日期、时间、纬度、经度、海拔。

五、程序示例

程序如下：

Main.C：	主程序
12864J：	液晶部分程序
GPS.C、GPS.H：	GPS 部分程序
CommInt：	接收字符，收到换行符(0X0A)，调用 GPS_Data，进行数据处理
CommInit：	串口初始化

GPS_CRC：　　　　　　　计算校验和
GPS_Data：　　　　　　 在 GPGGA 中提取海拔；在 GPRMC 中提取日期、时间、纬度、经度

六、思考题

1. 请将 GPS 接收到的时间转化为北京时间。

2. 本实验的 GPS 模块是静止不动的,如果 GPS 模块不是在空旷处或同时接收到信息的卫星数有限,测得的海拔误差较大。如果 GPS 模块安装在高速移动的物体上,如何根据接收到的信息,确定物体移动的速度、方向?

实验 14　GPRS 通信实验

一、实验目的

1. 了解短消息的编码、发送短消息格式、接收短消息格式。
2. 掌握 GB2312 编码转化为 UNICODE 编码方法。
3. 掌握使用 GPRS 模块发送短消息。

二、实验设备

STAR 系列实验仪一套、PC 机一台、GPRS 模块一块。

三、实验内容

1. 熟悉 GPRS 模块

① 本 GPRS 模块采用华为的 GTM900B 无线模块,它支持标准的 AT 命令及增强 AT 命令,提供丰富的数据业务等功能,内嵌 TCP/IP 协议、UDP 协议,支持 TEXT、PDU 等格式的短消息。

② 最高速率可达 85.6Kbps。通过 UART 接口与用户系统相连。

2. 短消息

短消息分为 TEXT、PDU 格式,TEXT 格式又分为 7 位编码、8 位编码。

(1) 7 位编码

如果发送的是英文信息,每个字符的最高位是零,可使用的 GSM 字符集为 7 位编码。设需要发送的短消息内容为"Hi",首先将字符转换为 7 位二进制,然后,将后面字符的位调用到前面,补齐前面的差别。例如,H 翻译成 1001000,i 翻译成 1101001,显然 H 的二进制编码不足 8 位,那么就将 i 的最后一位补足到 H 的前面,就成了 11001000(C8),i 剩下 6 位 110100,前面再补两个 0,变成 00110100(34),于是"Hi"就变成了两个八进制数 C8 和 34。

(2) PDU 数据格式

PDU 数据格式在短信正文中采用 UNICODE 编码,一个汉字由两个字节组成,通常汉字输入、显示采用 GB2312 编码格式,如果采用 PDU 格式发送短消息,发送前,将短消息正文改为 UNICODE 编码,否则,接收方会收到乱码。

(3) 短信中心号码

移动各地短信中心号码为1380+4位区号(不满4位,在区号后加0)+500。

(4) 发送短消息格式

例如,将字符"Hi"字符发送到目的地"13677328099",则PDU字符串为:
08 91 683108200105F0 11 00 0D 91 3176378290F9 00 00 00 02 C834

①08:短信息中心地址长度,指(91)+(683108200105F0)的长度。

②91:短信息中心号码类型。91是TON/NPI遵守International/E.164标准,指在号码前需加"+"号。此外还有其他数值,但91最常用。

91H:10010001B,十六进制91H等同于二进制数10010001B。其中每位表示的含义如下:

BIT No.	7	6	5	4	3	2	1	0
Name	1	数值类型			号码鉴别			

数值类型:000表示未知,001表示国际,010表示国内,111表示留作扩展;号码鉴别:0000表示未知,0001表示ISDN/电话号码(E.164/E.163),1111表示留作扩展。

③683108200105F0:短信息中心号码。由于位置上略有处理,实际号码应为:8613800210500(字母F是指长度减1)。这需要根据不同的地域作相应的修改。

上面①、②、③通称短消息中心地址(Address of the SMSC)。

④11:文件头字节。

BIT No.	7	6	5	4	3	2	1	0
Name	TP—RP	TP—UDHI	TP—SPR	TP—VFP		TP—RD	TP—MTI	
value	0	0	0	1	0	0	0	1

应答路径TP-RP(TP-Reply-Path):0表示不设置;1表示设置。

用户数据头标识TP-UDHL(TP-User-Data-Header-Indicator):0表示不含任何头信息;1表示含头信息。

状态报告要求TP-SPR(TP-Status-Report-Request):0表示需要报告;1表示不需要报告。

有效期格式TP-VPF(TP-Validity-Period-Format):00表示不提供(Not present);10表示整型(标准);01表示预留;11表示提供8位字节的一半(Semi—Octet Represented)。

拒绝复制TP-RD(TP-Reject-Duplicates):0表示接受复制;1表示拒绝复制。

信息类型提示TP-MTI(TP-Message-Type-Indicator):00表示读出(Deliver);01表示提交(Submit)。

⑤00:信息类型(TP-Message-Reference)。

⑥0B:被叫号码长度。

⑦91:被叫号码类型(同②)。

⑧3176378290F9:被叫号码,经过了位移处理,实际号码为"13677328099"。⑥、⑦、⑧通称"目的地址"(TP-Destination-Address)。

⑨00:协议标识TP-PID(TP-Protocol-Identifier)。

BIT No.7 6 5 4 3 2 1 0。

BIT No.7 与 BIT No.6：00—如下面定义的分配 BIT No.0—BIT No.5；01—参见 GSM03.40 协议标识完全定义；10 表示预留；11 表示为服务中心(SC)特殊用途分配。

BIT No.0—BIT No.5。

一般将这两位置为 00。

BIT No.5：0 表示不使用远程网络，只是短消息设备之间的协议；1 表示使用远程网络。

BIT No.0—Bits No.4：00000 表示隐含；00001 表示电传；00010 表示 group 3 telefax；00100 表示语音；00101 表示欧洲无线信息系统(ERMES)；00110 表示国内系统；10001 表示任何基于 X.400 的公用信息处理系统；10010 表示 Email。

⑩00：数据编码方案 TP-DCS(TP-Data-Coding-Scheme)。

BIT No.7 6 5 4 3 2 1 0。

BIT No.7 与 BIT No.6：一般设置为 00；BIT No.5：0 表示文本未压缩，1 表示文本用 GSM 标准压缩算法压缩；BIT No.4：0 表示 BIT No.1、BIT No.0 为保留位，不含信息类型信息，1 表示 BIT No.1、BIT No.0 含有信息类型信息；BIT No.3 与 BIT No.2：00 表示默认的字母表，01 表示 8bit，10 表示 USC2(16bit)，11 表示预留；BIT No.1 与 BIT No.0：00 表示 Class 0，01 表示 Class 1，10 表示 Class 2(SIM 卡特定信息)，11 表示 Class 3。

⑪00：有效期 TP-VP(TP-Valid-Period)。

VP value(&h)	相应的有效期
00 to 8F	(VP+1)*5 分钟
90 to A7	12 小时+(VP-143)*30 分钟
A8 to C4	(VP-166)*1 天
C5 to FF	(VP-192)*1 周

⑫02：用户数据长度 TP-UDL(TP-User-Data-Length)。

⑬C834：用户数据 TP-UD(TP-User-Data)"Hi"。

（5）发送短消息格式

①短信息中心地址长度（一个字节，同上）。

②短信息中心号码类型（一个字节，同上）。

③短信息中心号码。

④文件头字节（一个字节，同上）。

⑤被叫号码长度（一个字节，同上）。

⑥被叫号码类型（一个字节，同上）。

⑦被叫号码。

⑧协议标识 TP-PID（一个字节，同上）。

⑨数据编码方案 TP-DCS（一个字节，同上）。

⑩短信中心时间戳（7 个字节）。

⑪用户数据长度 TP-UDL（一个字节）。

⑫用户数据 TP-UD(TP-User-Data)：短信正文。

3. 相关的 GSM AT 指令

指令	功能
AT+CMGC	Send an SMS command(发出一条短消息命令)
AT+CMGD	Delete SMS message(删除 SIM 卡内存的短消息)
AT+CMGF	Select SMS message formate(选择短消息信息格式:0 表示 PDU;1 表示文本)
AT+CMGL	List SMS message from preferred store(列出 SIM 卡中的短消息 PDU/text:0/"REC UNREAD"表示未读,1/"REC READ"表示已读,2/"STO UNSENT"表示待发,3/"STO SENT"表示已发,4/"ALL"表示全部的)
AT+CMGR	Read SMS message(读短消息)
AT+CMGS	Send SMS message(发送短消息)
AT+CMGW	Write SMS message to memory(向 SIM 内存中写入待发的短消息)
AT+CMSS	Send SMS message from storage(从 SIM 内存中发送短消息)
AT+CNMI	New SMS message indications(显示新收到的短消息)
AT+CPMS	Preferred SMS message storage(选择短消息内存)
AT+CSCA	SMS service center address(短消息中心地址)
AT+CSCB	Select cell broadcast messages(选择蜂窝广播消息)
AT+CSMP	Set SMS text mode parameters(设置短消息文本模式参数)
AT+CSMS	Select Message Service(选择短消息服务)

4. 使用 GPRS 模块,发送一条英文短消息,接收一条英文短消息,发送一条中文短消息,接收一条中文短消息。

四、实验步骤

1. 连线说明:

将 GPRS 模块插入实验仪的 B5、C6 区。

A3 区:RXD、TXD	——	B5 区:KZ1_7、KZ1_6

2. 编写、调试程序,实现:发送一条英文短消息;接收一条英文短消息;发送一条中文短消息;接收一条中文短消息。

五、程序示例

程序如下:

```
Main.C:            主程序(循环收发短信)
comm.c:            串口初始化;接收 GPRS 发过来的数据;发送数据给 GPRS 模块
GPRS.C、GPRS.H:    GPRS 部分程序
```

1. 初始化短消息数据结构

void InitNoteStruct(u8 * pSMSC_TP, u8 * pTarget_TP);

pSMSC_TP:指向短信中心号码;pTarget_TP:指向手机号码

2. 发送短消息

bit SendNote(u8 * pNote, u8 length, bit bChinese);
pNote:指向短信正文;length:短信长度;bChinese:短信中带有汉字
3.接收短消息
bit ReceiveNotel(u8 * pNote, u8 * pLength, u8 * pTP, u8 * pTime);
pNote:指向存放短信正文存储器;pLength:存放短信长度;
pTP:指向发短信方手机号码;pTime:存放收到短信时间

七、思考题

两块 GPRS 模块之间如何使用 TCP/IP 协议收发数据？对方的 IP 号、端口号如何取得？

实验 15　非接触式卡实验

一、实验目的

1. 了解 MFRC500 读卡芯片特性、应用场合，了解非接触式卡工作原理。
2. 了解 MF1 IC S50 卡工作原理。
3. 掌握对 MF1 IC S50 卡读、写、增值、减值操作。

二、实验设备

STAR 系列实验仪一套、PC 机一台、MFRC500 模块一块、MF1 IC S50 卡一张。

三、实验内容

1. MFRC500 特性

①高集成度模拟电路用于卡应答的解调和解码。

②近距离操作(可达 100mm)，不需要增加有源电路。

③缓冲输出驱动器使用最少数目的外部元件连接到天线。

④载波频率为 13.56MHz。

⑤时钟频率监视。

⑥软件可实现掉电模式。

⑦并行微处理器接口带有内部地址锁存和 IRQ 线。

⑧自动检测微处理器并行接口类型。

⑨64 字节的发送和接收 FIFO 缓冲区。

⑩MF RC500 支持 ISO14443A 所有的层。

⑪支持 Crypto1 加密算法，提供可靠的内部非易失性密钥存储器。

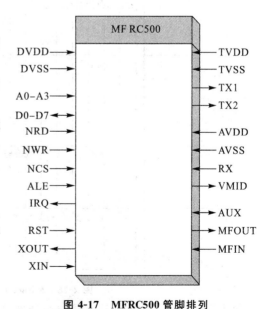

图 4-17　MFRC500 管脚排列

⑫支持 MIFARE Clasic。
⑬支持 MIRFARE 有源天线。
⑭内含可编程的定时器。
⑮支持防冲突过程。
⑯片内时钟电路,面向位和字节的帧。
⑰有防卡片重叠功能。
⑱唯一的序列号。
⑲内建 8 位/16 位的 CRC 协处理器,提供 CRC、PARITY 等数据校验。
⑳支持多种方式的活动天线,并且不需"天调系统"(天线调节系统)对天线进行补偿调节。

2. MF1 IC S50 卡

卡片上除了 IC 微晶片及一副高效率天线外,无任何其他元件;工作时的电源能量由卡片读写器天线发送无线电载波信号耦合到卡片上天线而产生电能,一般可达 2V 以上,供卡片上 IC 工作。工作频率为 13.56MHz。

(1)EEPROM

内含 1K 字节 EEPROM,分为 16 个扇区,每个扇区分为 4 个段,每个段有 16 个字节;可以自定义每个段的访问条件;数据可保持 10 年;可写 100,000 次。

Sector	Block	Byte Number within a Block															Description
		0	1	2	3	4	5	6	7	8	9	10	11	12	13	14	15	
15	3	Key A						Access Bits				Key B						Sector Trailer 15
	2																	Da
	1																	Da
	0																	Da
14	3	Key A						Access Bits				Key B						Sector Trailer 14
	2																	Da
	1																	Da
	0																	Da
⋮	⋮																	
1	3	Key A						Access Bits				Key B						Sector Trailer 1
	2																	Da
	1																	Da
	0																	Da
0	3	Key A						Access Bits				Key B						Sector Trailer 0
	2																	Da
	1																	Da
	0																	Manufacturer Blo

图 4-18 Mifare 1 卡片的存储结构

每个扇区的段 3(即第四段)是一个特殊的段,包含了该扇区的密码 A(6 个字节)、存取控制(4 个字节)、密码 B(6 个字节)。其余 3 个段是一般的数据段。

但扇区0的段0是特殊的,是厂商代码,已固化,不可改写。第0~4个字节为卡片的序列号,第5个字节为序列号的校验码;第6个字节为卡片的容量"SIZE"字节;第7、8个字节为卡片的类型号字节,即Tagtype字节;其他字节由厂商另加定义。

(2)安全性

安全性需要通过三轮确认(ISO/IEC DIS9798-2);RF信道的数据加密,有重防攻击保护;每个区有两套独立的密码;每张卡有唯一的序列号;卡片抗静电保护能力达2KV以上。

(3)操作

可执行读、写、增值、减值操作,如果需要增值/减值操作,必须对相应的段初始化。

下面将对密码A、密码B、存取控制与数据区的关系加以说明。

存取控制的结构如下:

位:7	6	5	4	3	2	1	0
C2X3_b	C2X2_b	C2X1_b	C2X0_b	C1X3_b	C1X2_b	C1X1_b	C1X0_b
C1X3	C1X2	C1X1	C1X0	C3X3_b	C3X2_b	C3X1_b	C3X0_b
C3X3	C3X2	C3X1	C3X0	C2X3	C2X2	C2X1	C2X0
BX7	BX6	BX5	BX4	BX3	BX2	BX1	BX0

_b表示取反,如C2X3_b表示C2X3取反;X表示扇区号;Y表示第几段;C表示控制位;B表示备用位。

存取控制对段3的控制如下(X=0~15):

C1X3	C2X3	C3X3	密码A read	密码A Write	存取控制 read	存取控制 write	密码B read	密码B write
0	0	0	never	KEYA|B	KEYA|B	never	KEYA|B	KEYA|B
0	1	0	never	Never	KEYA|B	never	KEYA|B	never
1	0	0	never	KEYB	KEYA|B	never	never	KEYB
1	1	0	never	Never	KEYA|B	never	never	never
0	0	1	never	KEYA|B	KEYA|B	KEYA|B	KEYA|B	KEYA|B
0	1	1	never	KEYB	KEYA|B	KEYB	never	KEYB
1	0	1	never	Never	KEYA|B	KEYB	never	never
1	1	1	never	Never	KEYA|B	never	never	never

KEYA|B表示密码A或密码B;never表示没有条件实现。

对数据段的控制如下表:(X＝0～15 扇区、Y＝每个扇区的 0～2 段)

C1XY	C2XY	C3XY	Read	Write	Increment	Decr,Transfer,restore
0	0	0	KEYA|B	KEYA|B	KEYA|B	KEYA|B
0	1	0	KEYA|B	never	Never	never
1	0	0	KEYA|B	KEYB	Never	never
1	1	0	KEYA|B	KEYB	KEYB	KEYA|B
0	0	1	KEYA|B	never	Never	KEYA|B
0	1	1	KEYB	KEYB	Never	never
1	0	1	KEYB	never	Never	never
1	1	1	Never	never	Never	never

块 3 的初始化值为:a0,a1,a2,a3,a4,a5,ff,07,80,69,b0,b1,b2,b3,b4,b5 共 16 个字节,其中 KEYA 是{a0,a1,a2,a3,a4,a5},KEYB 是{b0,b1,b2,b3,b4,b5},控制存取的 4 个字节为{0xff,0x07,0x80,0x69}。

根据具体的应用情况,对不同的扇区可选用不用的存取控制、不同的密码,但应注意其每一位的格式,以免误用!

数据段有两种应用方法:一种是用作一般的数据保存用,直接读写;另一种是用作数值段,可以进行初始化值、加值、减值、读值的运算。系统配用相应的函数完成相应的功能。

数值段的数据结构:

0	1	2	3	4	5	6	7	8	9	10	11	12	13	14	15
		VALUE						VALUE				Adr	Adr	Adr	Adr

数值段通过一个写操作,将存储的数据在一个段中写 3 次,1 次反写,从而完成数值段的初始化。此外,一个地址引导位代码域必须写 4 次,其中 2 次为反向写入。正/负数据值将以标准的 2 的补码格式来表示。

四、实验步骤

1. 连线说明:

将 MFRC500 模块插入实验仪的 B5 区。

B5 区:CZ1_2	——	A3 区:CS1
B5 区:CZ1_5	——	A3 区:INT0(P3.2)
B5 区:CZ1_7	——	A3 区:P1.0
D1 区:CTRL	——	A3 区:P1.1

2. 调用星研提供的 MFRC500.LIB 函数,实现对 MF1 IC S50 卡写、读、初始化数值段,对数值段加值、减值、读值操作。

五、程序示例

程序请参阅实验箱自带光盘 RC500 目录。